NONPARAMETRIC STATISTICS
FOR NON-STATISTICIANS

NONPARAMETRIC STATISTICS FOR NON-STATISTICIANS

A Step-by-Step Approach

GREGORY W. CORDER
DALE I. FOREMAN

A JOHN WILEY & SONS, INC., PUBLICATION

Copyright © 2009 by John Wiley & Sons, Inc. All rights reserved

Published by John Wiley & Sons, Inc., Hoboken, New Jersey
Published simultaneously in Canada

No part of this publication may be reproduced, stored in a retrieval system, or transmitted in any form or by any means, electronic, mechanical, photocopying, recording, scanning, or otherwise, except as permitted under Section 107 or 108 of the 1976 United States Copyright Act, without either the prior written permission of the Publisher, or authorization through payment of the appropriate per-copy fee to the Copyright Clearance Center, Inc., 222 Rosewood Drive, Danvers, MA 01923, (978) 750-8400, fax (978) 750-4470, or on the web at www.copyright.com. Requests to the Publisher for permission should be addressed to the Permissions Department, John Wiley & Sons, Inc., Ill River Street, Hoboken, NJ 07030, (201) 748-6011, fax (201) 748-6008, or online at http://www.wiley.com/go/permission.

Limit of Liability/Disclaimer of Warranty: While the publisher and author have used their best efforts in preparing this book, they make no representations or warranties with respect to the accuracy or completeness of the contents of this book and specifically disclaim any implied warranties of merchantability or fitness for a particular purpose. No warranty may be created or extended by sales representatives or written sales materials. The advice and strategies contained herein may not be suitable for your situation. You should consult with a professional where appropriate. Neither the publisher nor author shall be liable for any loss of profit or any other commercial damages, including but not limited to special, incidental, consequential, or other damages.

For general information on our other products and services or for technical support, please contact our Customer Care Department within the United States at (800) 762-2974, outside the United States at (317) 572-3993 or fax (317) 572-4002.

Wiley also publishes its books in a variety of electronic formats. Some content that appears in print may not be available in electronic formats. For more information about Wiley products, visit our web site at www.wiley.com.

Library of Congress Cataloging-in-Publication Data:

Corder, Gregory W., 1972-
  Nonparametric statistics for non-statisticians : a step-by-step approach /
Gregory W. Corder, Dale I. Foreman.
      p. cm.
  Includes bibliographical references and index.
  ISBN 978-0-470-45461-9 (cloth)
  1. Nonparametric statistics. I. Foreman, Dale I. II. Title.
  QA278.8.C67 2009
  519.5–dc22
                                                          2008049903

Printed in the United States of America

10 9 8 7 6 5 4 3 2

CONTENTS

Preface xi

1 Nonparametric Statistics: An Introduction 1

    1.1 Objectives 1
    1.2 Introduction 1
    1.3 The Nonparametric Statistical Procedures Presented in this Book 2
    1.4 Ranking Data 6
    1.5 Ranking Data with Tied Values 7
    1.6 Counts of Observations 8
    1.7 Summary 9
    1.8 Practice Questions 9
    1.9 Solutions to Practice Questions 10

2 Testing Data for Normality 12

    2.1 Objectives 12
    2.2 Introduction 12
    2.3 Describing Data and the Normal Distribution 13
    2.4 Computing and Testing Kurtosis and Skewness for Sample Normality 16
        2.4.1 Sample Problem for Examining Kurtosis 19
        2.4.2 Sample Problem for Examining Skewness 21
        2.4.3 Examining Skewness and Kurtosis for Normality Using SPSS® 23

	2.5	The Kolmogorov–Smirnov One-Sample Test	26
		2.5.1 Sample Kolmogorov–Smirnov One-Sample Test	28
		2.5.2 Performing the Kolmogorov–Smirnov One-Sample Test Using SPSS	32
	2.6	Summary	35
	2.7	Practice Questions	35
	2.8	Solutions to Practice Questions	36
3	**Comparing Two Related Samples: The Wilcoxon Signed Ranks Test**		**38**
	3.1	Objectives	38
	3.2	Introduction	38
	3.3	Computing the Wilcoxon Signed Ranks Test Statistic	39
		3.3.1 Sample Wilcoxon Signed Ranks Test (Small Data Samples)	40
		3.3.2 Performing the Wilcoxon Signed Ranks Test Using SPSS	42
		3.3.3 Confidence Interval for the Wilcoxon Signed Ranks Test	45
		3.3.4 Sample Wilcoxon Signed Ranks Test (Large Data Samples)	47
	3.4	Examples from the Literature	51
	3.5	Summary	51
	3.6	Practice Questions	52
	3.7	Solutions to Practice Questions	55
4	**Comparing Two Unrelated Samples: The Mann–Whitney U-Test**		**57**
	4.1	Objectives	57
	4.2	Introduction	57
	4.3	Computing the Mann–Whitney U-Test Statistic	58
		4.3.1 Sample Mann–Whitney U-Test (Small Data Samples)	59
		4.3.2 Performing the Mann–Whitney U-Test Using SPSS	62
		4.3.3 Confidence Interval for the Difference Between Two Location Parameters	66
		4.3.4 Sample Mann–Whitney U-Test (Large Data Samples)	68
	4.4	Examples from the Literature	72
	4.5	Summary	74
	4.6	Practice Questions	74
	4.7	Solutions to Practice Questions	77
5	**Comparing More Than Two Related Samples: The Friedman Test**		**79**
	5.1	Objectives	79
	5.2	Introduction	79
	5.3	Computing the Friedman Test Statistic	80
		5.3.1 Sample Friedman Test (Small Data Samples Without Ties)	81
		5.3.2 Sample Friedman Test (Small Data Samples with Ties)	84

		5.3.3	Performing the Friedman Test Using SPSS	88
		5.3.4	Sample Friedman Test (Large Data Samples Without Ties)	90
	5.4	Examples from the Literature		93
	5.5	Summary		95
	5.6	Practice Questions		95
	5.7	Solutions to Practice Questions		96

6 Comparing More than Two Unrelated Samples: The Kruskal–Wallis *H*-Test — 99

6.1	Objectives		99
6.2	Introduction		99
6.3	Computing the Kruskal–Wallis *H*-Test Statistic		100
	6.3.1	Sample Kruskal–Wallis *H*-Test (Small Data Samples)	101
	6.3.2	Performing the Kruskal–Wallis *H*-Test Using SPSS	106
	6.3.3	Sample Kruskal–Wallis *H*-Test (Large Data Samples)	110
6.4	Examples from the Literature		116
6.5	Summary		117
6.6	Practice Questions		117
6.7	Solutions to Practice Questions		118

7 Comparing Variables of Ordinal or Dichotomous Scales: Spearman Rank-Order, Point-Biserial, and Biserial Correlations — 122

7.1	Objectives		122
7.2	Introduction		122
7.3	The Correlation Coefficient		123
7.4	Computing the Spearman Rank-Order Correlation Coefficient		124
	7.4.1	Sample Spearman Rank-Order Correlation (Small Data Samples Without Ties)	125
	7.4.2	Sample Spearman Rank-Order Correlation (Small Data Samples with Ties)	128
	7.4.3	Performing the Spearman Rank-Order Correlation Using SPSS	131
7.5	Computing the Point-Biserial and Biserial Correlation Coefficients		134
	7.5.1	Correlation of a Dichotomous Variable and an Interval Scale Variable	134
	7.5.2	Correlation of a Dichotomous Variable and a Rank-Order Variable	136
	7.5.3	Sample Point-Biserial Correlation (Small Data Samples)	136
	7.5.4	Performing the Point-Biserial Correlation Using SPSS	139
	7.5.5	Sample Point-Biserial Correlation (Large Data Samples)	142
	7.5.6	Sample Biserial Correlation (Small Data Samples)	146
	7.5.7	Performing the Biserial Correlation Using SPSS	149

	7.6	Examples from the Literature	150
	7.7	Summary	151
	7.8	Practice Questions	151
	7.9	Solutions to Practice Questions	154

8 Tests for Nominal Scale Data: Chi-Square and Fisher Exact Test — 155

- 8.1 Objectives — 155
- 8.2 Introduction — 155
- 8.3 The Chi-Square Goodness-of-Fit Test — 156
  - 8.3.1 Computing the Chi-Square Goodness-of-Fit Test Statistic — 156
  - 8.3.2 Sample Chi-Square Goodness-of-Fit Test (Category Frequencies Equal) — 157
  - 8.3.3 Sample Chi-Square Goodness-of-Fit Test (Category Frequencies Not Equal) — 160
  - 8.3.4 Performing the Chi-Square Goodness-of-Fit Test Using SPSS — 163
- 8.4 The Chi-Square Test for Independence — 167
  - 8.4.1 Computing the Chi-Square Test for Independence — 168
  - 8.4.2 Sample Chi-Square Test for Independence — 169
  - 8.4.3 Performing the Chi-Square Test for Independence Using SPSS — 174
- 8.5 The Fisher Exact Test — 179
  - 8.5.1 Computing the Fisher Exact Test for 2×2 Tables — 180
  - 8.5.2 Sample Fisher Exact Test — 180
  - 8.5.3 Performing the Fisher Exact Test Using SPSS — 184
- 8.6 Examples from the Literature — 185
- 8.7 Summary — 187
- 8.8 Practice Questions — 187
- 8.9 Solutions to Practice Questions — 189

9 Test For Randomness: The Runs Test — 192

- 9.1 Objectives — 192
- 9.2 Introduction — 192
- 9.3 The Runs Test for Randomness — 193
  - 9.3.1 Sample Runs Test (Small Data Samples) — 194
  - 9.3.2 Performing the Runs Test Using SPSS — 195
  - 9.3.3 Sample Runs Test (Large Data Samples) — 200
  - 9.3.4 Sample Runs Test Referencing a Custom Value — 202
  - 9.3.5 Performing the Runs Test for a Custom Value Using SPSS — 204
- 9.4 Examples from the Literature — 208
- 9.5 Summary — 209

9.6	Practice Questions	209
9.7	Solutions to Practice Questions	210

Appendix A: SPSS at a Glance — 212

A.1	Introduction	212
A.2	Opening SPSS	212
A.3	Inputting Data	212
A.4	Analyzing Data	216
A.5	The SPSS Output	217

Appendix B: Tables of Critical Values — 219

Table B.1:	The Normal Distribution	219
Table B.2:	The Chi-Square Distribution	226
Table B.3:	Critical Values for the Wilcoxon Signed Ranks Test Statistics, T	227
Table B.4:	Critical Values for the Mann–Whitney U-Test Statistic	228
Table B.5:	Critical Values for the Friedman Test Statistic, F_r	230
Table B.6:	The Critical Values for the Kruskal–Wallis H-Test Statistic	231
Table B.7:	Critical Values for the Spearman Rank-Order Correlation Coefficient, r_s	238
Table B.8:	Critical Values for the Pearson Product-Moment Correlation Coefficient, r	239
Table B.9:	Factorials	241
Table B.10:	Critical Values for the Runs Test for Randomness	241

Bibliography — 243

Index — 245

PREFACE

The social and behavioral sciences need the ability to use nonparametric statistics in research. Many studies in these areas involve data that are classified on the nominal or ordinal scale. At times, interval data from these fields lack parameters for classification as normal. Nonparametric statistics is a useful tool for analyzing such data.

PURPOSE OF THIS BOOK

This book is intended to provide a conceptual and procedural approach to nonparametric statistics. It is written in such a manner as to enable someone who does not have an extensive mathematical background to work through the process necessary to conduct the given statistical tests presented. In addition, the outcome includes a discussion of the final decision for each statistical test. Each chapter takes the reader through an example from the beginning hypotheses, through the statistical calculations, to the final decision vis-a-vis the hypothesis taken into account. The examples are then followed by a detailed, step-by-step analysis by using the computer program, SPSS®. Finally, research literature is identified that uses the respective nonparametric statistical tests.

INTENDED AUDIENCE

Although not limited as such, this book is written for graduate and undergraduate students in social science programs. As stated earlier, it is targeted at the student who does not have an especially strong mathematical background and can be used effectively by students with both strong and weak mathematical backgrounds.

SPECIAL FEATURES OF THIS BOOK

There are currently few books available that provide a practical and applied approach to teaching nonparametric statistics. Many books take a more theoretical approach to the instructional process that can leave students disconnected and frustrated, in need of supplementary material to give them the ability to apply the statistics taught.

It is hoped that this book will provide students with a concrete approach to performing the nonparametric statistical procedures, along with their application and interpretation. We chose these particular nonparametric procedures since they represent a breadth of the typical analyses found in social science research. It is our hope that students will confidently learn the content presented with the promise of future successful applications.

In addition, each statistical test includes a section that explains how to use the computer program, SPSS. However, the organization of the book provides effective instruction of the nonparametric statistical procedures for those individuals with or without the software. Therefore, instructors (and students) can focus on learning the tests with a calculator, SPSS, or both.

A NOTE TO THE STUDENT

We have written this book with you in mind. Each of us has had a great deal of experience working with students just like you. It has been our experience that most people outside of the fields of mathematics or hard sciences struggle with and/or are intimidated by statistics. Moreover, we have found that when statistical procedures are explicitly communicated in a step-by-step manner, almost anyone can use them.

This book begins with a brief introduction (Chapter 1) and is followed with an explanation of how to perform the crucial step of checking your data for normality (Chapter 2). The chapters that follow (Chapters 3–9) highlight several nonparametric statistical procedures. Each of the chapters focuses on a particular type of variable and/or sample condition.

Chapters 3–9 have a similar organization. They each explain the statistical methods included in the respective chapters. At least one sample problem is included for each test using a step-by-step approach. (In some cases, we provide additional sample problems when procedures differ between large and small samples.) Then, those same sample problems are demonstrated by using the statistical software package, SPSS. Whether or not your instructor incorporates SPSS, this section will give you the opportunity to learn how to use the program. Toward the end of each chapter, we identify examples of the tests in published research. Finally, we present sample problems with solutions.

As you seek to learn nonparametric statistics, we strongly encourage you to work through the sample problems. Then, using the sample problems as a reference, work through the problems at the end of the chapters and additional data sets provided.

GREGORY W. CORDER

DALE I. FOREMAN

1

NONPARAMETRIC STATISTICS: AN INTRODUCTION

1.1 OBJECTIVES

In this chapter, you will learn the following items.

- The difference between parametric and nonparametric statistics.
- How to rank data.
- How to determine counts of observations.

1.2 INTRODUCTION

If you are using this book, it is possible that you have taken some type of introductory statistics class in the past. Most likely, your class began with a discussion about probability and later focused on particular methods for dealing with populations and samples. Correlations, z-scores, and t-tests were just some of the tools you might have used to describe populations and/or make inferences about a population using a simple random sample.

Many of the tests in a traditional, introductory statistics class are based on samples that follow certain assumptions called parameters. Such tests are called *parametric tests*. Specifically, parametric assumptions include samples that

---

Nonparametric Statistics for Non-Statisticians, Gregory W. Corder and Dale I. Foreman
Copyright © 2009 John Wiley & Sons, Inc.

- are randomly drawn from a normally distributed population,
- consist of independent observations, except for paired values,
- consist of values on an interval or ratio measurement scale,
- have respective populations of approximately equal variances,
- are adequately large,[1] and
- approximately resemble a normal distribution.

If any of your samples breaks one of these rules, you violate the assumptions of a parametric test. You do have some options, however.

You might change the nature of your study to adhere to the rules. If you are using an ordinal or nominal measurement scale, you might redesign your study to use an interval or ratio scale. (See Box 1.1 for a description of measurement scales.) Also, you may try to seek additional participants to enlarge your sample sizes. Unfortunately, there are times when one or neither of these changes is appropriate or even possible.

If your samples do not resemble a normal distribution, you might have learned to modify them so that you can use the tests you know. There are several legitimate ways to modify your data, so you can use parametric tests. First, if you can justify your reasons, you might remove extreme values from your samples called outliers. For example, imagine that you test a group of children and you wish to generalize the findings to typical children in a normal state of mind. After you collect the test results, most children earn scores around 80% with some scoring above and below the average. Suppose, however, that one child scored a 5%. If you find that this child does not speak English because he arrived in your country just yesterday, it would be reasonable to exclude his score from your analysis. Unfortunately, outlier removal is rarely this straightforward and deserves a much more lengthy discussion than we offer here.[2] Second, you can apply a mathematical adjustment to each value in your samples called a transformation. For example, you might square every value in a sample. Transformations do not always work, however. Third, there are more complicated methods that are beyond the scope of most introductory statistics courses. In such a case, you would be referred to a statistician.

Fortunately, there is a family of statistical tests that does not demand all the parameters, or rules, that we listed above. They are called *nonparametric tests* and this book will focus on several such tests.

1.3 THE NONPARAMETRIC STATISTICAL PROCEDURES PRESENTED IN THIS BOOK

This book describes several popular nonparametric statistical procedures used in research today. Table 1.1 identifies an overview of the types of tests presented in this book and their parametric counterparts.

---

[1] The minimum sample size for using a parametric statistical test varies among texts. For example, Pett (1997) and Salkind (2004) noted that most researchers suggest $n > 30$. Warner (2008) encouraged considering $n > 20$ as a minimum and $n > 10$ per group as an absolute minimum.

[2] Malthouse (2001) and Osborne and Overbay (2004) present discussions about the removal of outliers.

BOX 1.1 MEASUREMENT SCALES

Variables can be measured and conveyed in several ways. *Nominal* data, also called categorical data, are represented by counting the number of times a particular event or condition occurs. For example, you may categorize the political alignment of a group of voters. Group members could be labeled democratic, republican, independent, undecided, or other. No single person should fall into more than one category.

A *dichotomous* variable is a special classification of nominal data that is simply a measure of two conditions. A dichotomous variable is either discrete or continuous. A *discrete dichotomous* variable has no particular order and might include such examples as gender (male versus female) or a coin toss (heads versus tails). A *continuous dichotomous* variable has some type of order to the two conditions and might include measurements such as pass/fail or young/old.

Ordinal scale data describe values that occur in some order or rank. However, distance between any two ordinal values holds no particular meaning. For example, imagine lining up a group of people according to height. It would be very unlikely that the individual heights would increase evenly. Another example of an ordinal scale is a Likert-type scale. This scale asks the respondent to make a judgment using a scale with three, five, or seven items. The range of such a scale might use a 1 to represent strongly disagree, while a 5 might represent strongly agree. This type of scale can be considered an ordinal measurement since any two respondents will vary in their interpretations of the scale values.

An *interval* scale is a measure in which the relative distance between any two sequential values is the same. To borrow an example from physical science, consider the Celsius scale for measuring temperature. An increase from 3 degrees to 4 degrees is identical to an increase from 55 degrees to 56 degrees.

A *ratio* scale is slightly different from an interval scale. Unlike an interval scale, a ratio scale has an absolute zero value. In such a case, the zero value indicates a complete absence of a particular condition. To borrow another example from physical science, it would be appropriate to measure light intensity with a ratio scale. Total darkness is a complete absence of light and would receive a value of zero.

On a more general note, we have presented a classification of measurement scales similar to those used in many introductory statistics textbooks. To the best of our knowledge, this hierarchy of scales was first made popular by Stevens (1946). While Stevens has received agreement (Stake, 1960; Townsend & Ashby, 1984) and criticism (Anderson, 1961; Gaito, 1980; Velleman & Wilkinson, 1993), we believe the scale classifications we present suit the nature of this book. We direct anyone seeking additional information on this subject to the preceding citations.

TABLE 1.1

Type of Analysis	Nonparametric Test	Parametric Equivalent
Comparing two related samples	Wilcoxon signed ranks test	t-test for dependent samples
Comparing two unrelated samples	Mann–Whitney U-test	t-test for independent samples
Comparing three or more related samples	Friedman test	Repeated measures analysis of variance (ANOVA)
Comparing three or more unrelated samples	Kruskal–Wallis H-test	One-way analysis of variance (ANOVA)
Comparing categorical data	Chi-square tests and Fisher exact test	None
Comparing two rank-ordered variables	Spearman rank-order correlation	Pearson product-moment correlation
Comparing two variables when one variable is discrete dichotomous	Point-biserial correlation	Pearson product-moment correlation
Comparing two variables when one variable is continuous dichotomous	Biserial correlation	Pearson product-moment correlation
Examining a sample for randomness	Runs test	None

When demonstrating each nonparametric procedure, we will use a particular step-by-step method.

1. *State the null and research hypotheses.*

   First, we state the hypotheses for performing the test. The two types of hypotheses are null and alternate. The *null hypothesis* (H_0) is a statement that indicates no difference exists between conditions, groups, or variables. The *alternate hypothesis* (H_A), also called a research hypothesis, is the statement that predicts a difference or relationship between conditions, groups, or variables.

   The alternate hypothesis may be directional or nondirectional, depending on the context of the research. A directional, or one-tailed, hypothesis predicts a statistically significant change in a particular direction. For example, a treatment that predicts an improvement would be directional. A nondirectional, or two-tailed, hypothesis predicts a statistically significant change, but in no particular direction. For example, a researcher may compare two new conditions and predict a difference between them. However, he or she would not predict which condition would show the largest result.

2. *Set the level of risk (or the level of significance) associated with the null hypothesis.*

   When we perform a particular statistical test, there is always a chance that our result is due to chance instead of any real difference. For example, we might find that two samples are significantly different. Imagine, however, that no real difference exists. Our results would have led us to reject the null hypothesis when it was actually true. In this situation, we made a Type I error. Therefore, statistical tests assume some level of risk that we call alpha, or α.

There is also a chance that our statistical results would lead us to not reject the null hypothesis. However, if a real difference actually does exist, then we made a Type II error. We use the Greek letter beta, β, to represent a Type II error. See Table 1.2 for a summary of Type I and Type II errors.

TABLE 1.2

	We do not reject the null hypothesis	We reject the null hypothesis
The null hypothesis is actually true	No error	**Type I error**, α
The null hypothesis is actually false	**Type II error**, β	No error

After the hypotheses are stated, we choose the level of risk (or the level of significance) associated with the null hypothesis. We use the commonly accepted value of $\alpha = 0.05$. By using this value, there is a 95% chance that our statistical findings are real and not due to chance.

3. *Choose the appropriate test statistic.*

    We choose a particular type of test statistic based on characteristics of the data. For example, the number of samples or groups should be considered. Some tests are appropriate for two samples, while other tests are appropriate for three or more samples.

    Measurement scale also plays an important role in choosing an appropriate test statistic. We might select one set of tests for nominal data and a different set for ordinal variables. A common ordinal measure used in social and behavioral science research is the Likert scale. Nanna and Sawilowsky (1998) suggest that nonparametric tests are more appropriate for analyses involving Likert scales.

4. *Compute the test statistic.*

    The test statistic, or obtained value, is a computed value based on the particular test you need. Moreover, the method for determining the obtained value is described in each chapter and varies from test to test. For small samples, we use a procedure specific to a particular statistical test. For large samples, we approximate our data to a normal distribution and calculate a z-score for our data.

5. *Determine the value needed for rejection of the null hypothesis using the appropriate table of critical values for the particular statistic.*

    For small samples, we refer to a table of critical values provided in Appendix B. Each table provides a critical value to which we compare a computed test statistic. Finding a critical value using a table may require you to use such data characteristics as the degrees of freedom, number of measurements, and/or number of groups. In addition, you may need the desired level of risk, or alpha (α).

For large samples, we determine a critical region based on the level of risk (or the level of significance) associated with the null hypothesis, α. We will determine if the computed z-score falls within a critical region of the distribution.

6. *Compare the obtained value to the critical value.*

    Comparing the obtained value to the critical value allows us to identify a difference or relationship based on a particular level of risk. Once this is accomplished, we can state whether we must reject or must not reject the null hypothesis. While this type of phrasing may seem unusual, the standard practice in research is to state results in terms of the null hypothesis.

    Some of the tables of critical values are limited to particular sample or group size(s). When a sample size exceeds a table's range of value(s), we approximate our data to a normal distribution. In such cases, we use Table B.1 to establish a critical region of z-scores. Then, we calculate a z-score for our data and compare it to a critical region of z-scores. For example, if we use a two-tailed test with $\alpha = 0.05$, we do not reject the null hypothesis if $-1.96 \leq z \leq 1.96$.

7. *Interpret the results.*

    We can now give meaning to the numbers and values from our analysis based upon our context. If sample differences were observed, we can comment on the strength of those differences. We can compare observed results to expected results. We might examine a relationship between two variables for its relative strength or search a series of events for patterns.

8. *Reporting the results.*

    It may seem obvious that others cannot use our research unless we communicate our results in a meaningful and comprehensible manner. There is a fair amount of agreement in the research literature for reporting statistical results from parametric tests. Unfortunately, there is less agreement for nonparametric tests. We have attempted to use the more common reporting techniques found in the research literature.

1.4 RANKING DATA

Many of the nonparametric procedures involve ranking data values. Ranking values is really quite simple. Suppose that you are a math teacher and want to find out if students score higher after eating a healthy breakfast. You give a test and compare the scores of four students who ate a healthy breakfast with four students who did not. Table 1.3 shows the results.

TABLE 1.3

Students Who Ate Breakfast	Students Who Skipped Breakfast
87	93
96	83
92	79
84	73

TABLE 1.4

Value	Rank
73	1
79	2
83	3
84	4
87	5
92	6
93	7
96	8

To rank all of the values from Table 1.3 together, place them all in order in a new table from smallest to largest (see Table 1.4). The first value receives a rank of 1, the second value receives a rank of 2, and so on.

Notice that the values for the students who ate breakfast are in bold type. On the surface, it would appear that they scored higher. However, if you are seeking statistical significance, you need some type of procedure. The following chapters will offer those procedures.

1.5 RANKING DATA WITH TIED VALUES

The aforementioned ranking method should seem straightforward. In many cases, however, two or more of the data values may be repeated. We call repeated values *ties*, or tied values. Say, for instance, that you repeat the preceding ranking with a different group of students. This time, you collected new values shown in Table 1.5.

TABLE 1.5

Students Who Ate Breakfast	Students Who Skipped Breakfast
90	75
85	80
95	55
70	**90**

Rank the values as in the previous example. Notice that the value of 90 is repeated. This means that the value of 90 is a tie. If these two student scores were different, they would be ranked 6 and 7. In the case of a tie, give all of the tied values the average of their rank values. In this example, the average of 6 and 7 is 6.5 (see Table 1.6).

Most nonparametric statistical tests require a different formula when a sample of data contains ties. It is important to note that the formulas for ties are more algebraically complex. What is more, formulas for ties typically produce a test statistic that is only slightly different from the test statistic formulas for data without ties. It is probably for this reason that most statistics texts omit the formulas for tied

TABLE 1.6

Value	Rank Ignoring Tied Values	Rank Accounting for Tied Values
55	1	1
70	2	2
75	3	3
80	4	4
85	5	5
90	**6**	**6.5**
90	**7**	**6.5**
95	8	8

values. As you will see, however, we include the formulas for ties along with examples where applicable.

When the statistical tests in this book are explained using the computer program SPSS (Statistical Package for Social Scientists), there is no mention of any special treatment for ties. That is because SPSS automatically detects the presence of ties in any data set and applies the appropriate procedure for calculating the test statistic.

1.6 COUNTS OF OBSERVATIONS

Some nonparametric tests require *counts* (or frequencies) of observations. Determining the count is fairly straightforward and simply involves counting the total number of times a particular observations is made. For example, suppose you ask several children to pick their favorite ice cream flavor given three choices: vanilla, chocolate, and strawberry. Their preferences are shown in Table 1.7.

TABLE 1.7

Participant	Flavor
1	Chocolate
2	Chocolate
3	Vanilla
4	Vanilla
5	Strawberry
6	Chocolate
7	Chocolate
8	Vanilla

To find the counts for each ice cream flavor, list the choices and tally the total number of children who picked each flavor. In other words, count the number of children who picked chocolate. Then, repeat for the other choices, vanilla and strawberry. Table 1.8 reveals the counts from Table 1.7.

PRACTICE QUESTIONS

TABLE 1.8

Flavor	Count
Chocolate	4
Vanilla	3
Strawberry	1

To check your accuracy, you can add all the counts and compare them to the number of participants. The two numbers should be the same.

1.7 SUMMARY

In this chapter, we described differences between parametric and nonparametric tests. We also addressed assumptions by which nonparametric tests would be favorable over parametric tests. Then, we presented an overview of the nonparametric procedures included in this book. We also described the step-by-step approach we use to explain each test. Finally, we included explanations and examples of ranking and counting data, which are two tools for managing data when performing particular nonparametric tests.

The chapters that follow will present step-by-step directions for performing these statistical procedures both by manual, computational methods and by computer analysis using SPSS. In the next chapter, we address procedures for comparing data samples to a normal distribution.

1.8 PRACTICE QUESTIONS

1. Male high school students completed the one-mile run at the end of their ninth-grade year and the beginning of their tenth-grade year. The Table 1.9 values represent the differences between the recorded times. Notice that only one student's time improved (−2:08). Rank the values in Table 1.9 beginning with the student's time difference that displayed improvement.

TABLE 1.9

Participant	Value	Rank
1	0:36	
2	0:28	
3	1:41	
4	0:37	
5	1:01	
6	2:30	
7	0:44	
8	0:47	
9	0:13	
10	0:24	

(Continued)

Table 1.9 (*Continued*)

Participant	Value	Rank
11	0:51	
12	0:09	
13	−2:08	
14	0:12	
15	0:56	

2. The values in Table 1.10 represent weekly quiz scores on a math quiz. Rank the quiz scores.
3. Using the data from the previous example, what are the counts (or frequencies) of passing scores and failing scores if a 70 is a passing score?

TABLE 1.10

Participant	Score	Rank
1	100	
2	60	
3	70	
4	90	
5	80	
6	100	
7	80	
8	20	
9	100	
10	50	

1.9 SOLUTIONS TO PRACTICE QUESTIONS

1. The value ranks are listed in Table 1.11. Notice that there are no ties.

TABLE 1.11

Participant	Value	Rank
1	0:36	7
2	0:28	6
3	1:41	14
4	0:37	8
5	1:01	13
6	2:30	15
7	0:44	9
8	0:47	10
9	0:13	4
10	0:24	5
11	0:51	11
12	0:09	2
13	−2:08	1

SOLUTIONS TO PRACTICE QUESTIONS

Table 1.11 (*Continued*)

Participant	Value	Rank
14	0:12	3
15	0:56	12

2. The value ranks are listed in Table 1.12. Notice the tied values. The value of 80 occurred twice and required averaging the rank values of 5 and 6.

$$(5+6)/2 = 5.5$$

TABLE 1.12

Participant	Score	Rank
1	**100**	**9**
2	60	3
3	70	4
4	90	7
5	**80**	**5.5**
6	**100**	**9**
7	**80**	**5.5**
8	20	1
9	**100**	**9**
10	50	2

The value of 100 occurred three times and required averaging the rank values of 8, 9, and 10.

$$(8+9+10)/3 = 9$$

3. Table 1.13 shows the passing scores and failing scores using 70 as a passing score. Counts (or frequencies) of passing scores = 7 and failing scores = 3.

TABLE 1.13

Participant	Score	Pass/Fail
1	100	Pass
2	60	Fail
3	70	Pass
4	90	Pass
5	80	Pass
6	100	Pass
7	80	Pass
8	20	Fail
9	100	Pass
10	50	Fail

2

TESTING DATA FOR NORMALITY

2.1 OBJECTIVES

In this chapter, you will learn the following items.

- How to find a data sample's kurtosis and skewness and determine if the sample meets acceptable levels of normality.
- How to use SPSS to find a data sample's kurtosis and skewness and determine if the sample meets acceptable levels of normality.
- How to perform a Kolmogorov–Smirnov one-sample test to determine if a data sample meets acceptable levels of normality.
- How to use SPSS to perform a Kolmogorov–Smirnov one-sample test to determine if a data sample meets acceptable levels of normality.

2.2 INTRODUCTION

Parametric statistical tests, such as the t-test and one-way analysis of variance, are based on particular assumptions, or parameters. Those parameters are that data samples are randomly drawn from a normal population, consist of independent observations, are measured using an interval or ratio scale, are of an adequate size (see Chapter 1), and approximately resemble a normal distribution. Moreover, comparisons of samples or variables should have approximately equal variances.

Nonparametric Statistics for Non-Statisticians, Gregory W. Corder and Dale I. Foreman
Copyright © 2009 John Wiley & Sons, Inc.

DESCRIBING DATA AND THE NORMAL DISTRIBUTION

If data samples violate one or more of these assumptions, you should consider using a nonparametric test.

Examining the data gathering method, scale type, and size of a sample is fairly straightforward. However, examining a data sample's resemblance to a normal distribution, or its normality, requires a more involved analysis. Visually inspecting a graphical representation of a sample, such as a stem and leaf plot or a box and whisker plot, might be the most simplistic examination of normality. Statisticians advocate this technique in beginning statistics; however, this measure of normality does not suffice for strict levels of defensible analyses.

In this chapter, we present three quantitative measures of sample normality. First, we discuss the properties of the normal distribution. Then, we describe how to examine a sample's kurtosis and skewness. Next, we describe how to perform and interpret a Kolmogorov–Smirnov one-sample test. In addition, we describe how to perform each of these procedures using SPSS.

2.3 DESCRIBING DATA AND THE NORMAL DISTRIBUTION

An entire chapter could easily be devoted to the description of data and the normal distribution and many books do so. However, we will attempt to summarize the concept and begin with a practical approach as it applies to data collection.

In research, we often identify some population we wish to study. Then, we strive to collect several independent, random measurements of a particular variable associated with our population. We call this set of measurements a *sample*. If we use good experimental technique and our sample adequately represents our population, we can study the sample to make inferences about our population. For example, during a routine checkup, your physician draws a sample of your blood instead of all of your blood. This blood sample allows your physician to evaluate all of your blood even though he or she tested only the sample. Therefore, all of your body's blood cells represent the population about which your physician makes an inference using only the sample.

While a blood sample leads to the collection of a very large number of blood cells, other fields of study are limited to small sample sizes. It is not uncommon to collect less than 30 measurements for some studies in the behavioral and social sciences. Moreover, the measurements lie on some scale over which the measurements vary about the mean value. This notion is called *variance*. For example, a researcher uses some instrument to measure the intelligence of 25 children in a math class. It is highly unlikely that every child will have the same intelligence level. In fact, a good instrument for measuring intelligence should be sensitive enough to measure differences in the levels of the children.

The variance, s^2, can be expressed quantitatively. It can be calculated using Formula 2.1.

$$s^2 = \frac{\sum (x_i - \bar{x})^2}{n-1} \qquad (2.1)$$

where x_i is an individual value in the distribution, $\bar{x}$ is the distribution's mean, and n is the number of values in the distribution.

As mentioned in Chapter 1, parametric tests assume that the variances of samples being compared are approximately the same. This idea is called homogeneity of variance. To compare sample variances, Field (2005) suggested that we obtain a variance ratio by taking the largest sample variance and dividing it by the smallest sample variance. The variance ratio should be less than 2. Similarly, Pett (1997) indicated that no sample's variance be twice as large as any other sample's variance. If the homogeneity of variance assumption cannot be met, one would use a nonparametric test.

A more common way of expressing a sample's variability is with its standard deviation, s. Standard deviation is the square root of variance where $s = \sqrt{s^2}$. In other words, standard deviation is calculated using Formula 2.2.

$$s = \sqrt{\frac{\sum (x_i - \bar{x})^2}{n - 1}} \tag{2.2}$$

As illustrated in Figure 2.1, a small standard deviation indicates that a sample's values are fairly concentrated about mean, whereas a large standard deviation indicates that a sample's values are fairly spread out.

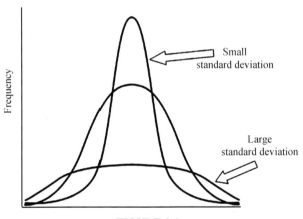

FIGURE 2.1

A histogram is a useful tool for graphically illustrating a sample's frequency distribution and variability (see Figure 2.2). This graph plots the value of the measurements horizontally and the frequency of each particular value vertically. The middle value is called the *median* and the greatest frequency is called the *mode*.

The mean and standard deviation of one distribution differs from the next. If we want to compare two or more samples, then we need some type of standard. A standard score is a way we can compare multiple distributions. The standard score that we use is called a z-score and it can be calculated using Formula 2.3.

$$z = \frac{x_i - \bar{x}}{s} \tag{2.3}$$

DESCRIBING DATA AND THE NORMAL DISTRIBUTION 15

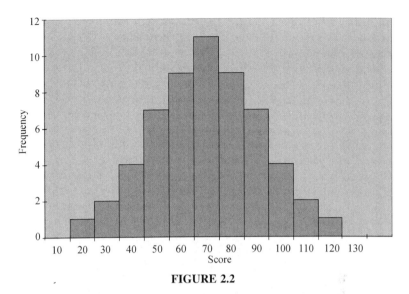

FIGURE 2.2

where x_i is an individual value in the distribution, $\bar{x}$ is the distribution's mean, and s is the distribution's standard deviation.

There is a useful relationship between the standard deviation and z-score. We can think of the standard deviation as a unit of horizontal distance away from the mean on the histogram. One standard deviation from the mean is the same as $z = 1.0$. Two standard deviations from the mean are the same as $z = 2.0$. For example, if $s = 10$ and $\bar{x} = 70$ for a distribution, then $z = 1.0$ at $x = 80$ and $z = 2.0$ at $x = 90$. What is more, z-scores that lie below the mean have negative values. Using our example, $z = -1.0$ at $x = 60$ and $z = -2.0$ at $x = 50$. Moreover, $z = 0.0$ at the mean value, $x = 70$. These z-scores can be used to compare our distribution to another distribution, even if the mean and standard deviation are different. In other words, we can compare multiple distributions in terms of z-scores.

To this point, we have been focused on distributions with finite numbers of values, n. As more data values are collected for a given distribution, the histogram begins to resemble a bell shape called the normal curve. Figure 2.3 shows the relationship

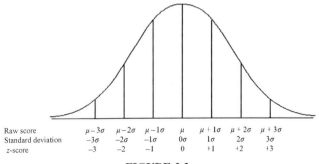

FIGURE 2.3

between the raw values, standard deviation, and z-scores of a population. Since we are describing a population, we use sigma, σ, to represent standard deviation and mu, μ, to represent the mean.

The normal curve has three particular properties (see Figure 2.4). First, the mean, median, and mode are equal. Thus, most of the values lie in the center of the distribution. Second, the curve displays perfect symmetry about the mean. Third, the left and right sides of the curve, called the tails, are asymptotic. This means that they approach the horizontal axis, but never touch it.

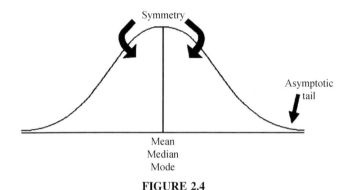

FIGURE 2.4

When we use a normal curve to represent probabilities, p, we refer to it as the normal distribution. We set the area under the curve equal to $p = 1.0$. Since the distribution is symmetrical about the mean, $p = 0.50$ on the left side of the mean and $p = 0.50$ on the right. In addition, the ordinate of the normal curve, y, is the height of the curve at a particular point. The ordinate is tallest at the curve's center and decreases as you move away from the center. Table B.1 provides the z-scores, probabilities, and ordinates for the normal distribution.

2.4 COMPUTING AND TESTING KURTOSIS AND SKEWNESS FOR SAMPLE NORMALITY

A frequency distribution that resembles a normal curve is approximately normal. However, not all frequency distributions have the approximate shape of a normal curve. The values might be densely concentrated in the center or substantially spread out. The shape of the curve may lack symmetry with many values concentrated on one side of the distribution. We use the terms kurtosis and skewness to describe these conditions, respectively.

Kurtosis is a measure of a sample or population that identifies how flat or peaked it is with respect to a normal distribution. Stated another way, kurtosis refers to how concentrated the values are in the center of the distribution. As shown in Figure 2.5, a peaked distribution is said to be leptokurtic. A leptokurtic distribution

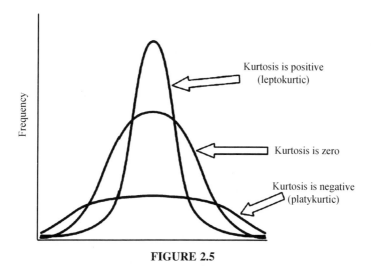

FIGURE 2.5

has a positive kurtosis. If a distribution is flat, it is said to be platykurtic. A platykurtic distribution has a negative kurtosis.

The skewness of a sample can be described as a measure of horizontal symmetry with respect to a normal distribution. As shown in Figure 2.6, if a distribution's scores are concentrated on the right side of the curve, it is said to be left skewed. A left skewed distribution has a negative skewness. If a distribution's scores are concentrated on the left side of the curve, it is said to be right skewed. A right skewed distribution has a positive skewness.

The kurtosis and skewness can be used to determine if a sample approximately resembles a normal distribution. There are five steps for examining sample normality in terms of kurtosis and skewness.

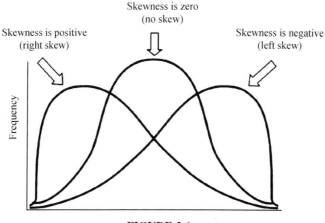

FIGURE 2.6

1. Determine the sample's mean and standard deviation.
2. Determine the sample's kurtosis and skewness.
3. Calculate the standard error of the kurtosis and the standard error of the skewness.
4. Calculate the z-score for the kurtosis and the z-score for the skewness.
5. Compare the z-scores to the critical region obtained from the normal distribution.

The calculations to find the values for a distribution's kurtosis and skewness require you to first find the sample mean, $\bar{x}$, and the sample standard deviation, s. Recall that standard deviation is found using Formula 2.2. The mean is found using Formula 2.4.

$$\bar{x} = \frac{\sum x_i}{n} \qquad (2.4)$$

where $\sum x_i$ is the sum of the values in the sample and n is the number of values in the sample.

The kurtosis, K, and standard error of the kurtosis, SE_K, are found using Formulas 2.5 and 2.6.

$$K = \left[\frac{n(n+1)}{(n-1)(n-2)(n-3)} \sum \left(\frac{x_i - \bar{x}}{s} \right)^4 \right] - \frac{3(n-1)^2}{(n-2)(n-3)} \qquad (2.5)$$

and

$$SE_K = \sqrt{\frac{24n(n-1)^2}{(n-2)(n-3)(n+5)(n+3)}} \qquad (2.6)$$

The skewness, S_k, and standard error of the skewness, SE_{S_k}, are found using Formulas 2.7 and 2.8.

$$S_k = \frac{n}{(n-1)(n-2)} \sum \left(\frac{x_i - \bar{x}}{s} \right)^3 \qquad (2.7)$$

and

$$SE_{S_k} = \sqrt{\frac{6n(n-1)}{(n-2)(n+1)(n+3)}} \qquad (2.8)$$

Normality can be evaluated using the z-score for the kurtosis, z_K, and the z-score for the skewness, z_{S_k}. Use Formulas 2.9 and 2.10 to find those z-scores.

$$z_K = \frac{K - 0}{SE_K} \qquad (2.9)$$

$$z_{S_k} = \frac{S_k - 0}{SE_{S_k}} \qquad (2.10)$$

TABLE 2.1

	Week 1 Quiz Scores	
90	72	90
64	95	89
74	88	100
77	57	35
100	64	95
65	80	84
90	100	76

Compare these z-scores to the values of the normal distribution (see Table B.1) for a desired level of confidence, α. For example, if you set $\alpha = 0.05$, then the calculated z-scores for an approximately normal distribution must fall between -1.96 and $+1.96$.

2.4.1 Sample Problem for Examining Kurtosis

The scores in Table 2.1 represent students' quiz performance during the first week of class. Use $\alpha = 0.05$ for your desired level of confidence. Determine if the sample of week 1 quiz scores is approximately normal in terms of its kurtosis.

First, find the mean of the sample.

$$\bar{x} = \frac{\sum x_i}{n} = \frac{1706}{21}$$
$$= 80.24$$

Next, find the standard deviation. It is helpful to set up Table 2.2 to manage the summation when computing the standard deviation (see Formula 2.2).

$$s = \sqrt{\frac{\sum(x_i - \bar{x})^2}{n-1}} = \sqrt{\frac{5525.81}{21-1}} = \sqrt{276.29}$$
$$= 16.62$$

Use the values for the mean and standard deviation to find the kurtosis. Again, it is helpful to set up Table 2.3 to manage the summation when computing the kurtosis (see Formula 2.5).

Compute the kurtosis.

$$K = \left[\frac{n(n+1)}{(n-1)(n-2)(n-3)} \sum \left(\frac{x_i - \bar{x}}{s}\right)^4\right] - \frac{3(n-1)^2}{(n-2)(n-3)}$$

$$= \left[\frac{21(21+1)}{(21-1)(21-2)(21-3)}(69.020)\right] - \frac{3(21-1)^2}{(21-2)(21-3)}$$

$$= \left[\frac{21(22)}{(20)(19)(18)}(69.020)\right] - \frac{3(20)^2}{(19)(18)}$$

$$= [0.0675(69.020)] - 3.509 = 4.662 - 3.509$$

$$= 1.153$$

TABLE 2.2

x_i	$x_i - \bar{x}$	$(x_i - \bar{x})^2$
90	9.76	95.29
72	−8.24	67.87
90	9.76	95.29
64	−16.24	263.68
95	14.76	217.91
89	8.76	76.77
74	−6.24	38.91
88	7.76	60.25
100	19.76	390.53
77	−3.24	10.49
57	−23.24	540.01
35	−45.24	2046.49
100	19.76	390.53
64	−16.24	263.68
95	14.76	217.91
65	−15.24	232.20
80	−0.24	0.06
84	3.76	14.15
90	9.76	95.29
100	19.76	390.53
76	−4.24	17.96
		$\sum(x_i - \bar{x})^2 = 5525.81$

Next, find the standard error of the kurtosis.

$$SE_K = \sqrt{\frac{24n(n-1)^2}{(n-2)(n-3)(n+5)(n+3)}}$$

$$= \sqrt{\frac{24(21)(21-1)^2}{(21-2)(21-3)(21+5)(21+3)}}$$

$$= \sqrt{\frac{24(21)(20)^2}{(19)(18)(26)(24)}} = \sqrt{\frac{201600}{213408}} = \sqrt{0.945}$$

$$= 0.972$$

Finally, use the kurtosis and the standard error of the kurtosis to find a z-score.

$$z_K = \frac{K - 0}{SE_K} = \frac{1.153 - 0}{0.972}$$

$$= 1.186$$

Use the z-score to examine the sample's approximation to a normal distribution. This value must fall between −1.96 and +1.96 to pass the normality assumption for

COMPUTING AND TESTING KURTOSIS AND SKEWNESS FOR SAMPLE NORMALITY

TABLE 2.3

x_i	$\frac{x_i - \bar{x}}{s}$	$\left(\frac{x_i - \bar{x}}{s}\right)^4$
90	0.587	0.119
72	−0.496	0.060
90	0.587	0.119
64	−0.977	0.911
95	0.888	0.622
89	0.527	0.077
74	−0.375	0.020
88	0.467	0.048
100	1.189	1.998
77	−0.195	0.001
57	−1.398	3.820
35	−2.722	54.864
100	1.189	1.998
64	−0.977	0.911
95	0.888	0.622
65	−0.917	0.706
80	−0.014	0.000
84	0.226	0.003
90	0.587	0.119
100	1.189	1.998
76	−0.255	0.004
		$\sum \left(\frac{x_i - \bar{x}}{s}\right)^4 = 69.020$

$\alpha = 0.05$. Since this z-score value does fall within that range, the sample has passed our normality assumption for kurtosis. Next, the sample's skewness must be checked for normality.

2.4.2 Sample Problem for Examining Skewness

Based on the same values from the example listed above, determine if the sample of week 1 quiz scores is approximately normal in terms of its skewness.

Use the mean and standard deviation from the previous example to find the skewness. Set up Table 2.4 to manage the summation in the skewness formula.

Compute the skewness.

$$S_k = \frac{n}{(n-1)(n-2)} \sum \left(\frac{x_i - \bar{x}}{s}\right)^3 = \frac{21}{(21-1)(21-2)}(-18.415)$$
$$= \frac{21}{(20)(19)}(-18.415)$$
$$= -1.018$$

TABLE 2.4

x_i	$\frac{x_i - \bar{x}}{s}$	$\left(\frac{x_i - \bar{x}}{s}\right)^3$
90	0.587	0.203
72	−0.496	−0.122
90	0.587	0.203
64	−0.977	−0.932
95	0.888	0.700
89	0.527	0.146
74	−0.375	−0.053
88	0.467	0.102
100	1.189	1.680
77	−0.195	−0.007
57	−1.398	−2.732
35	−2.722	−20.159
100	1.189	1.680
64	−0.977	−0.932
95	0.888	0.700
65	−0.917	−0.770
80	−0.014	0.000
84	0.226	0.012
90	0.587	0.203
100	1.189	1.680
76	−0.255	−0.017
		$\sum \left(\frac{x_i - \bar{x}}{s}\right)^3 = -18.415$

Next, find the standard error of the skewness.

$$SE_{S_k} = \sqrt{\frac{6n(n-1)}{(n-2)(n+1)(n+3)}} = \sqrt{\frac{6(21)(21-1)}{(21-2)(21+1)(21+3)}}$$

$$= \sqrt{\frac{6(21)(20)}{(19)(22)(24)}} = \sqrt{\frac{2520}{10032}} = \sqrt{0.251}$$

$$= 0.501$$

Finally, use the skewness and the standard error of the skewness to find a z-score.

$$z_{S_k} = \frac{S_k - 0}{SE_{S_k}} = \frac{-1.018}{0.501}$$

$$= -2.032$$

Use the z-score to examine the sample's approximation to a normal distribution. This value must fall between −1.96 and +1.96 to pass the normality assumption for $\alpha = 0.05$. Since this z-score value does not fall within that range, the sample has failed our normality assumption for skewness. Therefore, either the sample must be modified and rechecked, or you must use a nonparametric statistical test.

2.4.3 Examining Skewness and Kurtosis for Normality Using SPSS®

We will analyze the examples above using SPSS.

1. *Define your variables.*

   First, click the "Variable View" tab at the bottom of your screen. Then, type the name of your variable(s) in the "Name" column. As shown in Figure 2.7, we have named our variable "Wk1_Qz".

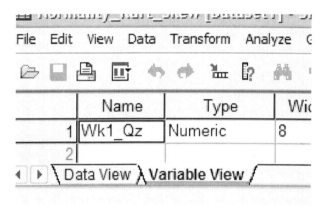

FIGURE 2.7

2. *Type in your values.*

   Click the "Data View" tab at the bottom of your screen and type your data under the variable names. As shown in Figure 2.8, we have typed the values for the "Wk1_Qz" sample.

	Wk1 Qz	var
1	90.00	
2	72.00	
3	90.00	
4	64.00	
5	95.00	
6	89.00	
7	74.00	
8	88.00	
9	100.00	
10	77.00	
11	57.00	
12	35.00	

FIGURE 2.8

3. *Analyze your data.*
   As shown in Figure 2.9, use the pull-down menus to choose "Analyze", "Descriptive Statistics", and "Descriptives...".

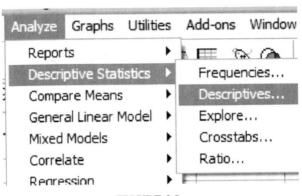

FIGURE 2.9

Choose the variable(s) that you want to examine. Then, click the button in the middle to move the variable to the "Variable(s)" box, as shown in Figure 2.10.

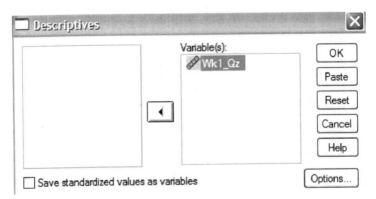

FIGURE 2.10

Next, click the "Options..." button to open the "Descriptives: Options" window shown in Figure 2.11. In the "Distribution" section, check the boxes next to "Kurtosis" and "Skewness". Then, click "Continue".

Finally, once you have returned to the "Descriptives" window, as shown in Figure 2.12, click "OK" to perform the analysis.

COMPUTING AND TESTING KURTOSIS AND SKEWNESS FOR SAMPLE NORMALITY 25

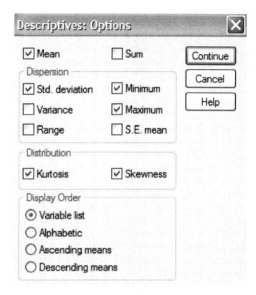

FIGURE 2.11

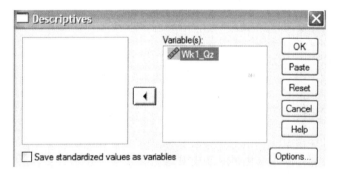

FIGURE 2.12

4. *Interpret the results from the SPSS Output window.*

Descriptive Statistics

	N	Minimum	Maximum	Mean	Std.	Skewness		Kurtosis	
	Statistic	Statistic	Statistic	Statistic	Statistic	Statistic	Std. Error	Statistic	Std. Error
Wk1_Qz	21	35.00	100.00	80.2381	16.62199	-1.018	.501	1.153	.972
Valid N (listwise)	21								

The SPSS Output provides the kurtosis and the skewness, along with their associated standard errors. In our example, the skewness is -1.018 and its standard error is 0.501. The kurtosis is 1.153 and its standard error is 0.972.

At this stage, we need to manually compute the z-scores for the skewness and kurtosis as we did in the previous examples. First, compute the z-score for kurtosis.

$$z_K = \frac{K - 0}{SE_K} = \frac{1.153 - 0}{0.972}$$
$$= 1.186$$

Next, we compute the z-score for skewness.

$$z_{S_k} = \frac{S_k - 0}{SE_{S_k}} = \frac{-1.018}{0.501}$$
$$= -2.032$$

Both of these values must fall between -1.96 and $+1.96$ to pass the normality assumption for $\alpha = 0.05$. The z-score for kurtosis falls within the desired range, but the z-score for skewness does not. Using $\alpha = 0.05$, the sample has passed the normality assumption for kurtosis, yet failed the normality assumption for skewness. Therefore, either the sample must be modified and rechecked, or you must use a nonparametric statistical test.

2.5 THE KOLMOGOROV–SMIRNOV ONE-SAMPLE TEST

The Kolmogorov–Smirnov one-sample test is a procedure to examine the agreement between two sets of values. For our purposes, the two sets of values compared are an observed frequency distribution based on a randomly collected sample and an empirical frequency distribution based on the sample's population. Furthermore, the observed sample is examined for normality when the empirical frequency distribution is based on a normal distribution.

The Kolmogorov–Smirnov test compares two cumulative frequency distributions. Using the point at which these two distributions show the largest divergence, the test identifies a two-tailed probability estimate, p, to determine if the samples are statistically similar or different.

To perform the Kolmogorov–Smirnov one-sample test, we begin by determining the relative empirical frequency distribution, $\hat{F}_{x_i}$, based on the observed sample. First, calculate the observed frequency distribution's midpoint, M, and standard deviation, s. The midpoint and standard deviation are found using Formulas 2.11 and 2.12.

$$M = (x_{\max} + x_{\min})/2 \qquad (2.11)$$

where $x_{\max}$ is the largest value in the sample and $x_{\min}$ is the smallest value in the sample.

$$s = \sqrt{\frac{\sum(f_i x_i^2) - \left[\left(\sum f_i x_i\right)^2 / n\right]}{n - 1}} \qquad (2.12)$$

where x_i is a given value in the observed sample, f_i is the frequency of a given value in the observed sample, and n is the number of values in the observed sample.

THE KOLMOGOROV–SMIRNOV ONE-SAMPLE TEST

Next, use the midpoint and standard deviation to calculate the z-scores (see Formula 2.13) for the sample values, x_i.

$$z = \left|\frac{x_i - M}{s}\right| \tag{2.13}$$

Use those z-scores and Table B.1 to determine the probability associated with each sample value, $\hat{p}_{x_i}$. These p-values are the relative frequencies of the empirical frequency distribution, $\hat{f}_r$.

Now, we find the relative values of the observed frequency distribution, f_r, using Formula 2.14.

$$f_r = \frac{f_i}{n} \tag{2.14}$$

where f_i is the frequency of a given value in the observed sample and n is the number of values in the observed sample.

Since the Kolmogorov–Smirnov test uses cumulative frequency distributions, both the relative empirical frequency distribution and relative observed frequency distribution must be converted into cumulative frequency distributions, $\hat{F}_{x_i}$ and S_{x_i}, respectively. Use Formulas 2.15 and 2.16 to find the absolute value divergence, $\tilde{D}$ and D, between the cumulative frequency distributions.

$$\tilde{D} = |\hat{F}_{x_i} - S_{x_i}| \tag{2.15}$$

$$D = |\hat{F}_{x_i} - S_{x_{i-1}}| \tag{2.16}$$

Use the largest divergence with Formula 2.17 to calculate the Kolmogorov–Smirnov test statistic, Z.

$$Z = \sqrt{n}\max(|D|, |\tilde{D}|) \tag{2.17}$$

Then, use the Kolmogorov–Smirnov test statistic, Z, and the Smirnov (1948) formula to find the two-tailed probability estimate, p.

$$\text{If } 0 \leq Z < 0.27, \quad \text{then } p = 1 \tag{2.18}$$

$$\text{If } 0.27 \leq Z < 1, \quad \text{then } p = 1 - \frac{2.506628}{Z}(Q + Q^9 + Q^{25}) \tag{2.19}$$

where

$$Q = e^{-1.233701Z^{-2}} \tag{2.20}$$

$$\text{If } 1 \leq Z < 3.1, \quad \text{then } p = 2(Q - Q^4 + Q^9 - Q^{16}) \tag{2.21}$$

where

$$Q = e^{-2Z^2} \tag{2.22}$$

$$\text{If } Z \geq 3.1, \quad \text{then } p = 0 \tag{2.23}$$

A *p*-value that exceeds the level of risk associated with the null hypothesis indicates that the observed sample is sufficiently normal for parametric statistics. Conversely, a *p*-value that is smaller than the level of risk indicates an observed sample is not sufficiently normal for parametric statistics. Such a lack of normality may require a nonparametric statistical test.

2.5.1 Sample Kolmogorov–Smirnov One-Sample Test

A department store has decided to evaluate customer satisfaction. As part of a pilot study, the store provides customers with a survey to rate employee friendliness. The survey uses a scale of 1–10 and its developer indicates that the scores should conform to a normal distribution. Use the Kolmogorov–Smirnov one-sample test to decide if the sample of customers surveyed responded with scores approximately matching a normal distribution. The survey results are shown in Table 2.5.

TABLE 2.5

Survey Results			
7	3	3	6
4	4	4	5
5	5	8	9
5	5	5	7
6	8	6	2

1. *State the null and research hypotheses.*

   The null hypothesis, shown below, states that the observed sample has an approximately normal distribution. The research hypothesis, shown below, states that the observed sample does not approximately resemble a normal distribution.

   The null hypothesis is

   H_0: There is no difference between the observed distribution of survey scores and a normally distributed empirical sample.

   The research hypothesis is

   H_A: There is a difference between the observed distribution of survey scores and a normally distributed empirical sample.

2. *Set the level of risk (or the level of significance) associated with the null hypothesis.*

   The level of risk, also called an alpha (α), is frequently set at 0.05. We will use an alpha of 0.05 in our example. In other words, there is a 95% chance that any observed statistical difference will be real and not due to chance.

3. *Choose the appropriate test statistic.*

   We are seeking to compare our observed sample against a normally distributed empirical sample. The Kolmogorov–Smirnov one-sample test will provide this comparison.

THE KOLMOGOROV–SMIRNOV ONE-SAMPLE TEST

4. *Compute the test statistic.*

   First, determine the midpoint and standard deviation for the observed sample. Table 2.6 helps to manage the summations for this process.

TABLE 2.6

Survey Score (x_i)	Score Frequency (f_i)	$f_i x_i$	$f_i x_i^2$
1	0	0	0
2	1	2	4
3	2	6	18
4	3	12	48
5	6	30	150
6	3	18	108
7	2	14	98
8	2	16	128
9	1	9	81
10	0	0	0
	$n = 20$	$\sum f_i x_i = 107$	$\sum f_i x_i^2 = 635$

Use Formula 2.11 to find the midpoint.

$$M = (x_{\max} + x_{\min})/2$$
$$= (9 + 2)/2$$
$$= 5.5$$

Then, use Formula 2.12 to find the standard deviation.

$$s = \sqrt{\frac{\sum(f_i x_i^2) - \left[(\sum f_i x_i)^2 / n\right]}{n - 1}}$$

$$= \sqrt{\frac{635 - (107^2/20)}{20 - 1}}$$

$$= 1.814$$

Now, determine the z-scores, empirical relative frequencies, and observed relative frequencies for each score value (see Table 2.7).

We will provide a sample calculation for survey score $= 4$ as seen in Table 2.7. Use Formula 2.13 to calculate the z-scores.

$$z = \left| \frac{x_i - M}{s} \right|$$

$$= \left| \frac{4 - 5.5}{1.814} \right|$$

$$= 0.827$$

TABLE 2.7

Survey Score (x_i)	Score Frequency (f_i)	z-Score	$\hat{p}_{x_i}$	Empirical Frequency ($\hat{f}_r$)	Observed Frequency (f_r)
1	0	2.398	0.0083	0.000	0.000
2	1	1.929	0.0269	0.027	0.050
3	2	1.378	0.0841	0.057	0.100
4	**3**	**0.827**	**0.2042**	**0.147**	**0.150**
5	6	0.276	0.3914	0.244	0.300
6	3	0.276	0.3914	0.244	0.150
7	2	0.827	0.2042	0.147	0.100
8	2	1.378	0.0841	0.057	0.100
9	1	1.929	0.0269	0.027	0.050
10	0	2.398	0.0083	0.000	0.000

Use that z-score and Table B.1 to determine the probability associated with the sample value, $\hat{p}_4$.

$$\hat{p}_4 = 0.2042$$

To find the empirical frequency value, $\hat{f}_{r4}$, for this p-value, subtract $\hat{f}_{r3}$ from $\hat{p}_4$. In other words,

$$\hat{f}_{r4} = \hat{p}_4 - \hat{f}_{r3}$$
$$= 0.2042 - 0.057$$
$$= 0.147$$

Formula 2.14 is used to calculate the relative values of the observed frequency distribution, f_r.

$$f_r = \frac{f_i}{n}$$
$$= \frac{3}{20}$$
$$= 0.150$$

Create cumulative distribution frequencies using the empirical and observed frequency distributions (see Table 2.8). Again, we provide a sample calculation for survey score = 4, as seen in bold in Table 2.8. Use Formulas 2.15 and 2.16 to find the absolute value divergence, $\tilde{D}$ and D, between the cumulative frequency distributions.

$$\tilde{D} = |\hat{F}_{x_i} - S_{x_i}|$$
$$= |0.231 - 0.300|$$
$$= 0.069$$

TABLE 2.8

Survey Score (x_i)	Relative Frequencies		Cumulative Frequencies		Cumulative Frequency Divergence	
	Empirical ($\hat{f}_r$)	Observed (f_r)	Empirical ($\hat{F}_{x_i}$)	Observed (S_{x_i})	$\tilde{D}$	D
1	0.000	0.000	0.000	0.000	0.000	
2	0.027	0.050	0.027	0.050	0.023	0.027
3	0.057	0.100	0.084	0.150	0.066	0.034
4	**0.147**	**0.150**	**0.231**	**0.300**	**0.069**	**0.081**
5	0.244	0.300	0.476	0.600	0.124	0.176*
6	0.244	0.150	0.720	0.750	0.030	0.120
7	0.147	0.100	0.867	0.850	0.017	0.117
8	0.057	0.100	0.924	0.950	0.026	0.074
9	0.027	0.050	0.951	1.000	0.049	0.001
10	0.000	0.000	0.951	1.000	0.049	0.049

and
$$D = |\hat{F}_{x_i} - S_{x_{i-1}}|$$
$$= |0.231 - 0.150|$$
$$= 0.081$$

To find the test statistic, Z, use the largest value from $\tilde{D}$ and D in Formula 2.17. The largest value is starred in Table 2.8, D = 0.176.

$$Z = \sqrt{n} \max(|D|, |\tilde{D}|)$$
$$= \sqrt{20}(0.176)$$
$$= 0.789$$

5. *Determine the p-value associated with the test statistic.*

The Kolmogorov–Smirnov test statistic, Z, and the Smirnov (1948) formula (see Formulas 2.18–2.23) are used to find the two-tailed probability estimate, *p*. Since $0.27 \leq Z < 1$, we use Formulas 2.19 and 2.20.

$$Q = e^{-1.23370 1Z^{-2}}$$
$$= e^{-1.233701(0.787)^{-2}}$$
$$= 0.1378$$

and

$$p = 1 - \frac{2.506628}{Z}(Q + Q^9 + Q^{25})$$
$$= 1 - \frac{2.506628}{0.789}(0.1378 + 0.1378^9 + 0.1378^{25})$$
$$= 0.562$$

6. *Compare the p-value to the level of risk (or the level of significance) associated with the null hypothesis.*

    The critical value for rejecting the null hypothesis is $\alpha = 0.05$ and the obtained p-value is $p = 0.562$. If the critical value is greater than the obtained value, we must reject the null hypothesis. If the critical value is less than the obtained p-value, we must not reject the null hypothesis. Since the critical value is less than the obtained value ($0.05 < 0.562$), we do not reject the null hypothesis.

7. *Interpret the results.*

    We did not reject the null hypothesis, suggesting the customers' survey ratings of employee friendliness sufficiently resembled a normal distribution. This means that a parametric statistical procedure may be used with this sample.

8. *Reporting the results.*

    When reporting the results from the Kolmogorov–Smirnov one-sample test, we include the test statistic (D), the degrees of freedom (which equals the sample size), and the p-value in terms of the level of risk, α. Based on our analysis, the sample of customers is approximately normal where $D_{(20)} = 0.176$, $p > 0.05$.

2.5.2 Performing the Kolmogorov–Smirnov One-Sample Test Using SPSS

We will analyze the data from the example above using SPSS.

1. *Define your variables.*

    First, click the "Variable View" tab at the bottom of your screen. Then, type the names of your variables in the "Name" column. As shown in Figure 2.13, the variable is called "Survey".

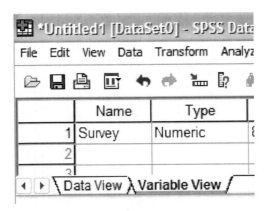

FIGURE 2.13

2. *Type in your values.*
   Click the "Data View" tab at the bottom of your screen. Type your sample values in the "Survey" column as shown in Figure 2.14.

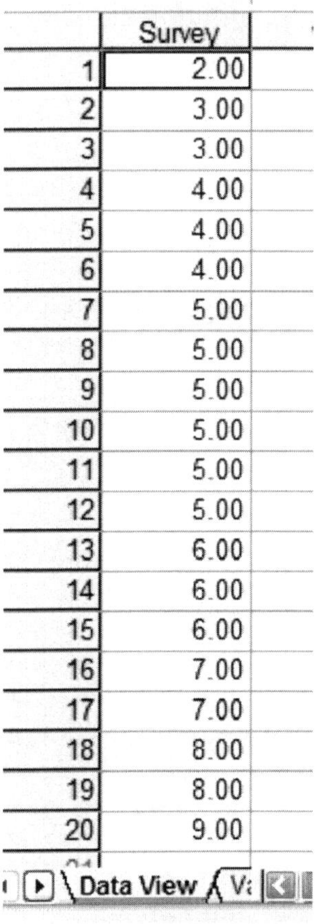

	Survey
1	2.00
2	3.00
3	3.00
4	4.00
5	4.00
6	4.00
7	5.00
8	5.00
9	5.00
10	5.00
11	5.00
12	5.00
13	6.00
14	6.00
15	6.00
16	7.00
17	7.00
18	8.00
19	8.00
20	9.00

FIGURE 2.14

3. *Analyze your data.*
   As shown in Figure 2.15, use the pull-down menus to choose "Analyze", "Nonparametric Tests", and "1-Sample K-S...".

   Use the arrow button to place your variable with your data values in the box labeled "Test Variable List:" as shown in Figure 2.16. Finally, click "OK" to perform the analysis.

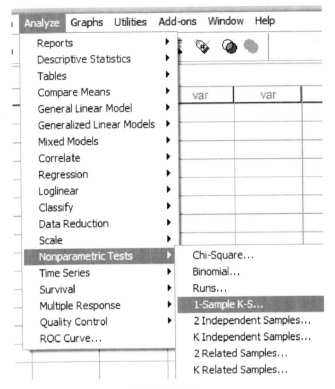

FIGURE 2.15

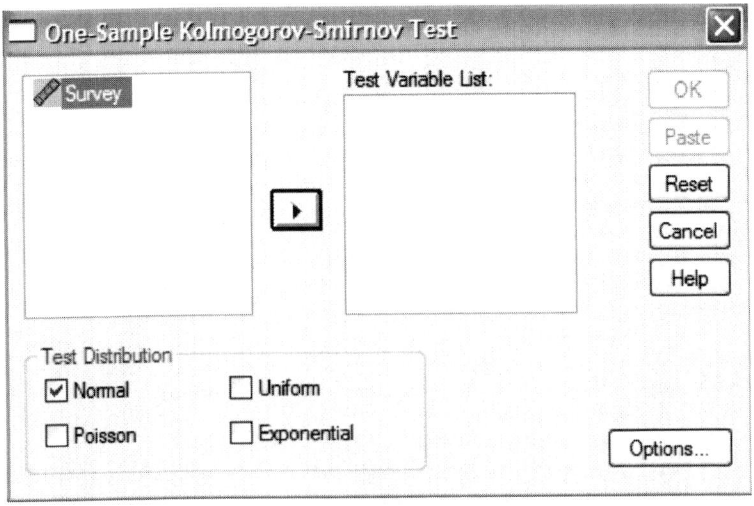

FIGURE 2.16

4. *Interpret the results from the SPSS Output window.*

One-Sample Kolmogorov-Smirnov Test

		Survey
N		20
Normal Parameters[a,b]	Mean	5.3500
	Std. Deviation	1.81442
Most Extreme Differences	Absolute	.176
	Positive	.176
	Negative	-.124
Kolmogorov-Smirnov Z		.789
Asymp. Sig. (2-tailed)		.562

a. Test distribution is Normal.
b. Calculated from data.

The SPSS Output provides the most extreme difference ($D=0.176$), Kolmogorov–Smirnov Z-test statistic ($Z=0.789$), and the significance ($p=0.562$). Based on the results from SPSS, the p-value exceeds the level of risk associated with the null hypothesis ($\alpha=0.05$). Therefore, we do not reject the null hypothesis. In other words, the sample distribution is sufficiently normal.

2.6 SUMMARY

Parametric statistical tests, such as the t-test and one-way analysis of variance, are based on particular assumptions, or parameters. Therefore, it is important that you examine collected data for its approximation to a normal distribution. Upon doing that, you can consider whether you will use a parametric or nonparametric test for analyzing your data.

In this chapter, we presented three quantitative measures of sample normality. First, we described how to examine a sample's kurtosis and skewness. Then, we described how to perform and interpret a Kolmogorov–Smirnov one-sample test. In the following chapters, we will describe several nonparametric procedures for analyzing data samples that do not meet the assumptions needed for parametric statistical tests. In the next chapter, we will begin by describing a test for comparing two unrelated samples.

2.7 PRACTICE QUESTIONS

1. The values in Table 2.9 are a sample of reading-level score for a ninth-grade class. They are measured on a ratio scale. Examine the sample's skewness and kurtosis for normality for $\alpha=0.05$. Report your findings.

TABLE 2.9

Ninth-Grade Reading-Level Scores									
8.10	8.20	8.20	8.70	8.70	8.80	8.80	8.90	8.90	8.90
9.20	9.20	9.20	9.30	9.30	9.30	9.40	9.40	9.40	9.40
9.50	9.50	9.50	9.50	9.60	9.60	9.60	9.70	9.70	9.90

2. Using a Kolmogorov–Smirnov one-sample test, examine the sample of values from Table 2.9. Report your findings.

2.8 SOLUTIONS TO PRACTICE QUESTIONS

1. SPSS returned the following values:
   Skewness $= -0.904$
   Standard error of the skewness $= 0.427$
   Kurtosis $= 0.188$
   Standard error of the kurtosis $= 0.833$
   The computed z-scores are below.

   $$z_{S_k} = -2.117$$

   and

   $$z_K = 0.226$$

   At $\alpha = 0.05$, the sample's skewness fails the normality test, while the kurtosis passes the normality test. Based on our standard of $\alpha = 0.05$, this sample of reading levels for ninth-grade students is not sufficiently normal.

2. The SPSS Output shows the results from the Kolmogorov–Smirnov one-sample test.

One-Sample Kolmogorov-Smirnov Test

			Scores
N			30
Normal Parameters[a,b]	Mean		9.1800
	Std. Deviation		.46639
Most Extreme Differences	Absolute		.184
	Positive		.099
	Negative		-.184
Kolmogorov-Smirnov Z			1.007
Asymp. Sig. (2-tailed)			.263

a. Test distribution is Normal.
b. Calculated from data.

Kolmogorov–Smirnov obtained value = 1.007
Two-tailed significance = 0.263

According to the Kolmogorov–Smirnov one-sample test with $\alpha = 0.05$, this sample of reading levels for ninth-grade students is sufficiently normal.

3

COMPARING TWO RELATED SAMPLES: THE WILCOXON SIGNED RANKS TEST

3.1 OBJECTIVES

In this chapter, you will learn the following items.

- How to compute the Wilcoxon signed ranks test.
- How to perform the Wilcoxon signed ranks test using SPSS.
- How to construct a median confidence interval based on the Wilcoxon signed ranks test for matched pairs.

3.2 INTRODUCTION

Imagine that you give an attitude test to a small group of people. After you deliver some type of treatment, say a daily vitamin C supplement for several weeks, you give that same group of people another attitude test. Finally, you compare the two measures of attitude to see if there is any type of difference between the two sets of scores.

The two sets of test scores in the previous scenario are related, or paired. This is because each person was tested twice. In other words, each test score in one group of scores has another test score counterpart. The Wilcoxon signed ranks test is a nonparametric statistical procedure for comparing two samples that are paired, or related. The parametric equivalent to the Wilcoxon signed ranks test goes by names

---

Nonparametric Statistics for Non-Statisticians, Gregory W. Corder and Dale I. Foreman
Copyright © 2009 John Wiley & Sons, Inc.

such as the Student's *t*-test, *t*-test for matched pairs, *t*-test for paired samples, or *t*-test for dependent samples.

In this chapter, we will describe how to perform and interpret a Wilcoxon signed ranks test for both small samples and large samples. We will also explain how to perform the procedure using SPSS. Finally, we offer varied examples of these nonparametric statistics from the literature.

3.3 COMPUTING THE WILCOXON SIGNED RANKS TEST STATISTIC

The formula for computing the Wilcoxon T for small samples is shown in Formula 3.1. The signed ranks are the values that are used to compute the positive and negative values in the formula.

$$T = \text{smaller of } \sum R_+ \text{ and } \sum R_- \qquad (3.1)$$

where ΣR_+ is the sum of the ranks with positive differences and ΣR_- is the sum of the ranks with negative differences.

After the T statistic is computed, it must be examined for significance. We may use a table of critical values (see Table B.3). However, if the number of pairs, n, exceeds those available from the table, then a large sample approximation may be performed. For large samples, compute a z-score and use a table with the normal distribution (see Table B.1) to obtain a critical region of z-scores. Formulas 3.2–3.4 are used to find the z-score of a Wilcoxon signed ranks test for large samples.

$$\bar{x}_T = \frac{n(n+1)}{4} \qquad (3.2)$$

where $\bar{x}_T$ is the mean and n is the number of matched pairs included in the analysis.

$$s_T = \sqrt{\frac{n(n+1)(2n+1)}{24}} \qquad (3.3)$$

where s_T is the standard deviation.

$$z^* = \frac{T - \bar{x}_T}{s_T} \qquad (3.4)$$

where z^* is the z-score for an approximation of the data to the normal distribution and T is the T statistic.

At this point, the analysis is limited to identifying the presence or absence of a significant difference between the groups and does not describe the strength of the treatment. We can consider the effect size (ES) to determine the degree of association between the groups. We use Formula 3.5 to calculate the effect size.

$$ES = \frac{|z|}{\sqrt{n}} \qquad (3.5)$$

where $|z|$ is the absolute value of the z-score and n is the number of matched pairs included in the analysis.

The effect size ranges from 0 to 1. Cohen (1988) defined the conventions for effect size as small = 0.10, medium = 0.30, and large = 0.50. (Correlation coefficient and effect size are both measures of association. See Chapter 7 concerning correlation for more information on Cohen's assignment of effect size's relative strength.)

3.3.1 Sample Wilcoxon Signed Ranks Test (Small Data Samples)

The counseling staff of Clear Creek County School District has implemented a new program this year to reduce bullying in their elementary schools. The school district does not know if the new program resulted in improvement or deterioration. To evaluate the program's effectiveness, the school district has decided to compare the percent of successful interventions last year before the program began with the percent of successful interventions this year with the program in place. In Table 3.1, the 12 elementary school counselors, or participants, reported the percent of successful interventions last year and the percent this year.

TABLE 3.1

	Percent of Successful Interventions	
Participant	Last Year	This Year
1	31	31
2	14	14
3	53	50
4	18	30
5	21	28
6	44	48
7	12	35
8	36	32
9	22	23
10	29	34
11	17	27
12	40	42

The samples are relatively small, so we need a nonparametric procedure. Since we are comparing two related, or paired, samples, we will use the Wilcoxon signed ranks test.

1. *State the null and research hypotheses.*

    The null hypothesis states that the counselors reported no difference in the percentages last year and this year. The research hypothesis states that the counselors observed some differences between this year and last year. Our research hypothesis is a two-tailed, nondirectional hypothesis because it indicates a difference, but in no particular direction.

    The null hypothesis is
    $H_0: \mu_D = 0$

COMPUTING THE WILCOXON SIGNED RANKS TEST STATISTIC

The research hypothesis is
$$H_A: \quad \mu_D \neq 0$$

2. *Set the level of risk (or the level of significance) associated with the null hypothesis.*

   The level of risk, also called an alpha (α), is frequently set at 0.05. We will use an alpha of 0.05 in our example. In other words, there is a 95% chance that any observed statistical difference will be real and not due to chance.

3. *Choose the appropriate test statistic.*

   The data are obtained from 12 counselors, or participants, who are using a new program designed to reduce bullying among students in the elementary schools. The participants reported the percent of successful interventions last year and the percent this year. We are comparing last year's percentages with this year's percentages. Therefore, the data samples are related, or paired. In addition, sample sizes are relatively small. Since we are comparing two related samples, we will use the Wilcoxon signed ranks test.

4. *Compute the test statistic.*

   First, compute the difference between each sample pair. Then, rank the absolute value of those computed differences. Using this method, the differences of zero are ignored when ranking. We have done this in Table 3.2.

TABLE 3.2

Participant	Percent of Successful Interventions		Difference	Rank Without Zero	Sign
	Last Year	This Year			
1	31	31	0	Exclude	
2	14	14	0	Exclude	
3	53	50	−3	3	−
4	18	30	+12	9	+
5	21	28	+7	7	+
6	44	48	+4	4.5	+
7	12	35	+23	10	+
8	36	32	−4	4.5	−
9	22	23	+1	1	+
10	29	34	+5	6	+
11	17	27	+10	8	+
12	40	42	+2	2	+

Compute the sum of ranks with positive differences. Using Table 3.2, the ranks with positive differences are 9, 7, 4.5, 10, 1, 6, 8, and 2. When we add all of the ranks with positive difference, we get $\sum R_+ = 47.5$.

Compute the sum of ranks with negative differences. The ranks with negative differences are 3 and 4.5. The sum of ranks with negative difference is $\sum R_- = 7.5$.

The obtained value is the smaller of the two rank sums. Therefore, the Wilcoxon is $T = 7.5$.

5. *Determine the value needed for rejection of the null hypothesis using the appropriate table of critical values for the particular statistic.*

   Since the sample sizes are small, we use Table B.3, which lists the critical values for the Wilcoxon T. As noted in Table 3.2, the two counselors with score differences of zero were discarded. This reduces our sample size to $n = 10$. In this case, we look for the critical value under the two-tailed test for $n = 10$ and $\alpha = 0.05$. Table B.3 returns a critical value for the Wilcoxon test of $T = 8$. An obtained value that is less than or equal to 8 will lead us to reject our null hypothesis.

6. *Compare the obtained value to the critical value.*

   The critical value for rejecting the null hypothesis is 8 and the obtained value is $T = 7.5$. If the critical value equals or exceeds the obtained value, we must reject the null hypothesis. If instead, the critical value is less than the obtained value, we must not reject the null hypothesis. Since the critical value exceeds the obtained value, we must reject the null hypothesis.

7. *Interpret the results.*

   We rejected the null hypothesis, suggesting that a real difference exists between last year's percentages and this year's percentages. In addition, since the sum of the positive difference ranks ($\sum R_+$) was larger than the negative difference ranks ($\sum R_-$), the difference is positive, showing a positive impact of the program. Therefore, our analysis provides evidence that the new bullying program is providing positive benefits toward the improvement of student behavior as perceived by the school counselors.

8. *Reporting the results.*

   When reporting the findings, include the T statistic, sample size, and p-value's relation to α. The directionality of the difference should be expressed using the sum of the positive difference ranks ($\sum R_+$) and sum of the negative difference ranks ($\sum R_-$).

   For this example, the Wilcoxon signed ranks test ($T = 7.5$, $n = 12$, $p < 0.05$) indicated that the percentage of successful interventions was significantly different. In addition, the sum of the positive difference ranks ($\sum R_+ = 47.5$) was larger than the sum of the negative difference ranks ($\sum R_- = 7.5$), showing a positive impact from the program. Therefore, our analysis provides evidence that the new bullying program is providing positive benefits toward the improvement of student behavior as perceived by the school counselors.

3.3.2 Performing the Wilcoxon Signed Ranks Test Using SPSS

We will analyze the above example using SPSS.

1. *Define your variables.*

   First, click the "Variable View" tab at the bottom of your screen. Then, type the names of your variables in the "Name" column. As shown in Figure 3.1, we have named our variables "last_yr" and "this_yr".

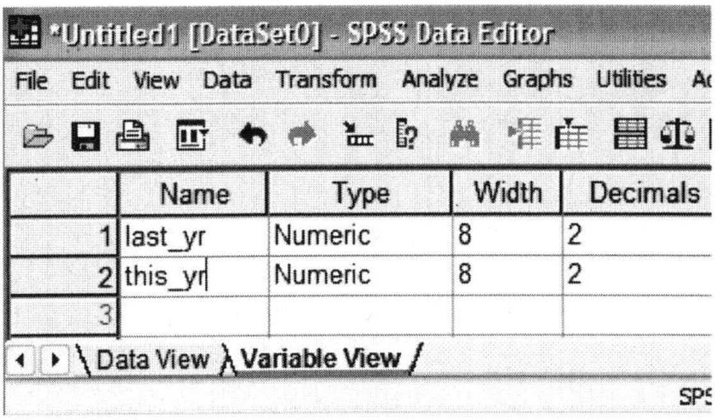

COMPUTING THE WILCOXON SIGNED RANKS TEST STATISTIC

FIGURE 3.1

2. *Type in your values.*
   Click the "Data View" tab at the bottom of your screen and type your data under the variable names. As shown in Figure 3.2, we are comparing "last_yr" with "this_yr".

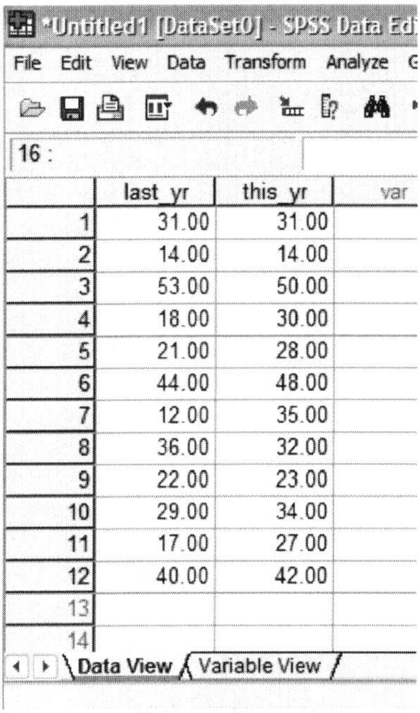

	last_yr	this_yr
1	31.00	31.00
2	14.00	14.00
3	53.00	50.00
4	18.00	30.00
5	21.00	28.00
6	44.00	48.00
7	12.00	35.00
8	36.00	32.00
9	22.00	23.00
10	29.00	34.00
11	17.00	27.00
12	40.00	42.00

FIGURE 3.2

3. *Analyze your data.*
   As shown in Figure 3.3, use the pull-down menus to choose "Analyze", "Nonparametric Tests", and "2 Related Samples...".

COMPARING TWO RELATED SAMPLES: THE WILCOXON SIGNED RANKS TEST

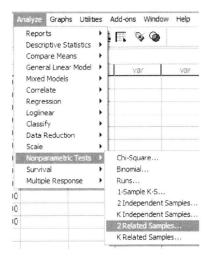

FIGURE 3.3

In the upper left box, select both variables that you want to compare. Then, use the arrow button to place your variable pair in the box labeled "Test Variable List". After the variable pair is in the "Test Pair(s) List" (see Figure 3.4), click "OK" to perform the analysis.

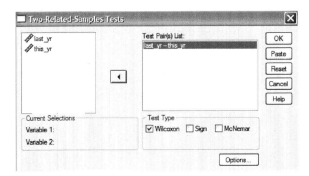

FIGURE 3.4

4. *Interpret the results from the SPSS Output window.*

Ranks

		N	Mean Rank	Sum of Ranks
this_yr - last_yr	Negative Ranks	2ª	3.75	7.50
	Positive Ranks	8ᵇ	5.94	47.50
	Ties	2ᶜ		
	Total	12		

a. this_yr < last_yr
b. this_yr > last_yr
c. this_yr = last_yr

The first SPSS output table provides the Wilcoxon T, or obtained value. From the "Sum of Ranks" column, choose the smaller of the two values. In our example, $T = 7.5$.

Test Statistics[b]

	this_yr - last_yr
Z	-2.040[a]
Asymp. Sig. (2-tailed)	.041

a. Based on negative ranks.
b. Wilcoxon Signed Ranks Test

SPSS also returns the critical z-score for large samples. In addition, SPSS calculates the two-tailed significance (0.041).

Based on the results from SPSS, the percentage of successful interventions was significantly different ($T = 7.5$, $n = 12$, $p < 0.05$). In addition, the sum of the positive difference ranks ($\sum R_+ = 47.5$) was larger than the sum of the negative difference ranks ($\sum R_- = 7.5$), demonstrating a positive impact from the program.

Note: Using the above procedure, SPSS always returns the two-tailed significance. If you were to need a one-tailed significance, simply divide by 2. The one-tailed significance in the above example is $0.041/2 = 0.0205$.

3.3.3 Confidence Interval for the Wilcoxon Signed Ranks Test

The American Psychological Association (2001) has suggested that researchers report the *confidence interval* for research data. A confidence interval is an inference to a population in terms of an estimation of sampling error. More specifically, it provides a range of values that fall within the population with a level of confidence of $100(1 - \alpha)\%$.

A median confidence interval can be constructed based on the Wilcoxon signed ranks test for matched pairs. To create this confidence interval, all of the possible matched pairs (X_i, X_j) are used to compute the differences $D_i = X_i - X_j$. Then, compute all of the averages, u_{ij}, of two difference scores using Formula 3.6. There will be a total of $[n(n-1)/2] + n$ averages.

$$u_{ij} = (D_i + D_j)/2, \qquad 1 \leq i \leq j \leq n \qquad (3.6)$$

We will perform a 95% confidence interval using the sample Wilcoxon signed ranks test with a small data sample (see above). Table 3.1 provides the values for obtaining our confidence interval. We begin by using Formula 3.6 to compute all of the averages, u_{ij}, of two difference scores. For example,

$$u_{11} = (D_1 + D_1)/2 = (-3 + -3)/2$$
$$= -3$$
$$u_{12} = (D_1 + D_2)/2 = (-3 + 12)/2$$
$$= 4.5$$
$$u_{13} = (D_1 + D_3)/2 = (-3 + 7)/2$$
$$= 2$$

Table 3.3 shows each value of u_{ij}.

TABLE 3.3

	−3	12	7	4	23	−4	1	5	10	2
−3	−3	4.5	2	0.5	10	−3.5	−1	1	3.5	−0.5
12		12	9.5	8	17.5	4	6.5	8.5	11	7
7			7	5.5	15	1.5	4	6	8.5	4.5
4				4	13.5	0	2.5	4.5	7	3
23					23	9.5	12	14	16.5	12.5
−4						−4	−1.5	0.5	3	−1
1							1	3	5.5	1.5
5								5	7.5	3.5
10									10	6
2										2

Next, arrange all of the averages in order from smallest to largest. We have arranged all of the values for u_{ij} in Table 3.4.

TABLE 3.4

1	−4.0	12	1.0	22	4.0	34	6.5	45	10.0
2	−3.5	13	1.5	23	4.0	35	7.0	46	11.0
3	−3.0	14	1.5	24	4.0	36	7.0	47	12.0
4	−1.5	15	2.0	25	4.5	37	7.0	48	12.0
5	−1.0	15	2.0	26	4.5	38	7.5	49	12.5
6	−1.0	16	2.5	27	4.5	39	8.0	50	13.5
7	−0.5	17	3.0	28	5.0	40	8.5	51	14.0
8	0.0	18	3.0	29	5.5	41	8.5	52	15.0
9	0.5	19	3.0	30	5.5	42	9.5	53	16.5
10	0.5	20	3.5	31	6.0	43	9.5	54	17.5
11	1.0	21	3.5	32	6.0	44	10.0	55	23.0

The median of the ordered averages gives a point estimate of the population median difference. The median of this distribution is 4.5, which is the point estimate of the population.

Use Table B.3 to find the end points of the confidence interval. First, determine T from the table that corresponds with the sample size and desired confidence such that $p = \alpha/2$. We seek to find a 95% confidence interval. For our example, $n = 10$ and $p = 0.05/2$. The table provides $T = 8$.

The end points of the confidence interval are the Kth smallest and the Kth largest values of u_{ij}, where $K = T + 1$. For our example, $K = 8 + 1 = 9$. The ninth value from the bottom is 0.5 and the ninth value from the top is 12.0. Based on these findings, it is estimated with 95% confidence that the difference of successful interventions due to the new bullying programs lies between 0.5 and 12.0.

3.3.4 Sample Wilcoxon Signed Ranks Test (Large Data Samples)

Hearing of Clear Creek School District's success with their antibullying program, Jonestown School District has implemented the program this year to reduce bullying in their own elementary schools. The Jonestown School District evaluates their program's effectiveness by comparing the percent of successful interventions last year before the program began with the percent of successful interventions this year with the program in place. In Table 3.5, the 25 elementary school counselors, or participants, reported the percent of successful interventions last year and the percent this year.

TABLE 3.5

Participant	Percent of Successful Interventions	
	Last Year	This Year
1	53	50
2	18	43
3	21	28
4	44	48
5	12	35
6	36	32
7	22	23
8	29	34
9	17	27
10	10	42
11	38	44
12	37	16
13	19	33
14	37	50
15	28	20
16	15	27
17	25	27
18	38	30
19	40	51
20	30	50
21	23	45
22	41	20
23	31	49
24	28	43
25	14	30

COMPARING TWO RELATED SAMPLES: THE WILCOXON SIGNED RANKS TEST

We will use the same nonparametric procedure to analyze the data. However, we use a large sample ($n \geq 20$) approximation.

1. *State the null and research hypotheses.*

    The null hypothesis states that the counselors reported no difference in the percentages last year and this year. The research hypothesis states that the counselors observed some differences between this year and last year. Our research hypothesis is a two-tailed, nondirectional hypothesis because it indicates a difference, but in no particular direction.

    The null hypothesis is

    $H_0: \mu_D \neq 0$

    The research hypothesis is

    $H_A: \mu_D \neq 0$

2. *Set the level of risk (or the level of significance) associated with the null hypothesis.*

    The level of risk, also called an alpha (α), is frequently set at 0.05. We will use an alpha of 0.05 in our example. In other words, there is a 95% chance that any observed statistical difference will be real and not due to chance.

3. *Choose the appropriate test statistic.*

    The data are obtained from 25 counselors, or participants, who are using a new program designed to reduce bullying among students in the elementary schools. The participants reported the percent of successful interventions last year and the percent this year. We are comparing last year's percentages with this year's percentages. Therefore, the data samples are related, or paired. Since we are comparing two related samples, we will use the Wilcoxon signed ranks test.

4. *Compute the test statistic.*

    First, compute the difference between each sample pair. Then, rank the absolute value of those computed differences. We have done this in Table 3.6.

    Compute the sum of ranks with positive differences. Using Table 3.6, when we add all of the ranks with positive difference we get $\sum R_+ = 257.5$.

    Compute the sum of ranks with negative differences. The ranks with negative differences are 3, 4.5, 9.5, 9.5, 20.5, and 20.5. The sum of ranks with negative difference is $\sum R_- = 67.5$. The obtained value is the smaller of these two rank sums. Thus, the Wilcoxon $T = 67.5$.

    Since our sample size is larger than 20, we will approximate it to a normal distribution. Therefore, we will find a z-score for our data using a normal approximation. We must find the mean, $\bar{x}_T$, and the standard deviation, s_T, for the data.

$$\bar{x}_T = \frac{n(n+1)}{4} = \frac{25(25+1)}{4}$$
$$= 162.5$$

COMPUTING THE WILCOXON SIGNED RANKS TEST STATISTIC

TABLE 3.6

Participant	Percent of Successful Interventions		Difference	Rank	Sign
	Last Year	This Year			
1	53	50	−3	3	−
2	18	43	+25	24	+
3	21	28	+7	8	+
4	44	48	+4	4.5	+
5	12	35	+23	23	+
6	36	32	−4	4.5	−
7	22	23	+1	1	+
8	29	34	+5	6	+
9	17	27	+10	11	+
10	10	42	+32	25	+
11	38	44	+6	7	+
12	37	16	−21	20.5	−
13	19	33	+14	15	+
14	37	50	+13	14	+
15	28	20	−8	9.5	−
16	15	27	+12	13	+
17	25	27	+2	2	+
18	38	30	−8	9.5	−
19	40	51	+11	12	+
20	30	50	+20	19	+
21	23	45	+22	22	+
22	41	20	−21	20.5	−
23	31	49	+18	18	+
24	28	43	+15	16	+
25	14	30	+16	17	+

and

$$s_T = \sqrt{\frac{n(n+1)(2n+1)}{24}} = \sqrt{\frac{25(25+1)(50+1)}{24}} = \sqrt{\frac{33150}{24}}$$
$$= 37.17$$

Next, we use the mean, standard deviation, and the T test statistic to calculate a z-score. Remember, we are testing the hypothesis that there is no difference in ranks of percentages of successful interventions between last year and this year.

$$z^* = \frac{T - \bar{x}_T}{s_T} = \frac{67.5 - 162.5}{37.17}$$
$$= -2.56$$

5. *Determine the value needed for rejection of the null hypothesis using the appropriate table of critical values for the particular statistic.*

   Table B.1 is used to establish the critical region of z-scores. For a two-tailed test with $\alpha = 0.05$, we must not reject the null hypothesis if $-1.96 \leq z^* \leq 1.96$.

6. *Compare the obtained value to the critical value.*

   We find that z^* is not within the critical region of the distribution, $-2.56 < -1.96$. Therefore, we reject the null hypothesis. This suggests a difference in the percentage of successful interventions after the program was implemented.

7. *Interpret the results.*

   We rejected the null hypothesis, suggesting that a real difference exists between last year's percentages and this year's percentages. In addition, since the sum of the positive difference ranks ($\sum R_+$) was larger than the negative difference ranks ($\sum R_-$), the difference is positive, showing a positive impact of the program. Therefore, our analysis provides evidence that the new bullying program is providing positive benefits toward the improvement of student behavior as perceived by the school counselors.

   At this point, the analysis is limited to identifying the presence or absence of a significant difference between the groups. In other words, the statistical test's level of significance does not describe the strength of the treatment. The American Psychological Association (2001), however, has called for a measure of the strength called the *effect size*.

   We can consider the effect size for this large sample test to determine the degree of association between the groups. We use Formula 3.5 to calculate the effect size. For the example, $|z| = 2.56$ and $n = 25$.

$$\text{ES} = \frac{|z|}{\sqrt{n}} = \frac{|-2.56|}{\sqrt{25}}$$
$$= 0.51$$

   Our effect size for the matched-pair samples is 0.51. This value indicates a high level of association between the percent of successful interventions before and after the implementation of the new bullying program.

8. *Reporting the results.*

   For this example, the Wilcoxon signed ranks test ($T = 67.5, n = 25, p < 0.05$) indicated that the percentage of successful interventions was significantly different. In addition, the sum of the positive difference ranks ($\sum R_+ = 257.5$) was larger than the sum of the negative difference ranks ($\sum R_- = 67.5$), showing a positive impact from the program. Moreover, the effect size for the matched-pair samples was 0.51. Therefore, our analysis provides evidence that the new bullying program is providing positive benefits toward the improvement of student behavior as perceived by the school counselors.

3.4 EXAMPLES FROM THE LITERATURE

Below are varied examples of the nonparametric procedures described in this chapter. We have summarized each study's research problem and the researchers' rationale(s) for choosing a nonparametric approach. We encourage you to obtain these studies if you are interested in their results.

- Boser, J., & Poppen, W. A. (1978). Identification of teacher verbal response roles for improving student–teacher relationships. *Journal of Educational Research*, 72(2), 91–94.

    Boser and Poppen sought to determine which verbal responses by the teacher held the greatest potential for improving student–teacher relationships. The seven verbal responses were feelings, thoughts, motives, behaviors, encounter/encouragement, confrontation, and sharing. They used a Wilcoxon signed ranks test to examine 101 ninth-grader responses because the student participants rank ordered their responses.

- Vaughn, S., Reiss, M., Rothlein, L., & Hughes, M. T. (1999). Kindergarten teachers' perceptions of instructing students with disabilities. *Remedial and Special Education*, 20(3), 184–191.

    Vaughn, Reiss, Rothlein, and Hughes investigated kindergarten teachers' perceptions of practices identified to improve outcomes for children with disabilities transitioning from prekindergarten to kindergarten. The researchers compared the paired ratings of teachers' desirability to employ the identified practices with feasibility using a Wilcoxon signed ranks test. This nonparametric procedure was considered the most appropriate because the study's measure was a Likert-type scale (1 = low, 5 = high).

- Rinderknecht, K., & Smith, C. (2004). Social Cognitive Theory in an after-school nutrition intervention for urban Native American youth. *Journal of Nutrition Education & Behavior*, 36(6), 298–304.

    Rinderknecht and Smith used a 7-month nutrition intervention to improve the dietary self-efficacy of Native American children (5–10 years) and adolescents (11–18 years). Wilcoxon signed ranks tests were used to determine whether fat and sugar intake changed significantly between pre- and postintervention among adolescents. The researchers chose nonparametric tests for their data that were not normally distributed.

3.5 SUMMARY

Two samples that are paired, or related, may be compared using a nonparametric procedure called the Wilcoxon signed ranks test. The parametric equivalent to this test is known as the Student's t-test, t-test for matched pairs, or t-test for dependent samples.

3.6 PRACTICE QUESTIONS

1. A teacher wished to determine if providing a bilingual dictionary to students with limited English proficiency improves math test scores. A small class of students ($n = 10$) was selected. Students were given two math tests. Each test covered the same type of math content; however, students were provided a bilingual dictionary on the second test. The data in Table 3.7 represents the students' performance on each math test.

   Use a one-tailed Wilcoxon signed ranks test with $\alpha = 0.05$ to determine which testing condition resulted in higher scores. Report your findings.

TABLE 3.7

Student	Math Test Without a Bilingual Dictionary	Math Test with a Bilingual Dictionary
1	30	39
2	56	46
3	48	37
4	47	44
5	43	32
6	45	39
7	36	41
8	44	40
9	44	38
10	40	46

2. A research study was done to investigate the influence of being alone at night on the human male heart rate. Ten men were sent into a wooded area, one at a time, at night, for 20 min. They had a heart monitor to record their pulse rate. The second night, the same men were sent into a similar wooded area accompanied by a companion. Their pulse rate was recorded again. The researcher wanted to see if having a companion would change their pulse rate. The median rates are reported in Table 3.8.

   Use a two-tailed Wilcoxon signed ranks test with $\alpha = 0.05$ to determine which condition produced a higher pulse rate. Report your findings.

TABLE 3.8

Participant	Median Rate Alone	Median Rate with Companion
A	88	72
B	77	74
C	91	80
D	70	77
E	80	71
F	85	83
G	90	80
H	82	91
I	93	86
J	75	69

3. A researcher conducts a pilot study to compare two treatments to help obese female teenagers lose weight. She tests each individual in two different treatment conditions. The data in Table 3.9 provides the number of pounds that each participant lost.

   Use a two-tailed Wilcoxon signed ranks test with $\alpha = 0.05$ to determine which treatment resulted in greater weight loss. Report your findings.

TABLE 3.9

	Pounds Lost	
Participant	Treatment 1	Treatment 2
1	10	18
2	20	12
3	15	16
4	9	7
5	18	21
6	11	17
7	6	13
8	12	14

4. Twenty participants in a exercise program were measured on the number of sit-ups they could do before other physical exercise (first count) and the number they could do after they had done at least 45 minutes of other physical exercise (second count). Table 3.10 shows the results for 20 participants obtained during two separate physical exercise sessions. Determine the effect size for a calculated z-score.

TABLE 3.10

Participant	First Count	Second Count
1	18	28
2	19	18
3	20	28
4	29	20
5	15	30
6	22	25
7	21	28
8	30	18
9	22	27
10	11	30
11	20	24
12	21	27
13	21	10
14	20	40
15	18	20
16	27	14
17	24	29
18	13	30
19	10	24
20	10	36

5. A school is trying to get more students to participate in activities that will make learning more desirable. Table 3.11 shows the number of activities that each of the 10 students in one class participated in last year before a new activity program was implemented and this year after it was implemented. Construct a 95% median confidence interval based on the Wilcoxon signed ranks test to determine whether the new activity program had a significant positive effect on the student participation.

TABLE 3.11

Participant	Last Year	This Year
1	18	20
2	22	28
3	10	18
4	25	23
5	16	20
6	14	21
7	21	17
8	13	18
9	28	22
10	12	21

3.7 SOLUTIONS TO PRACTICE QUESTIONS

1. The results from the analysis are displayed in the SPSS Outputs.

Ranks

		N	Mean Rank	Sum of Ranks
with_D - without_D	Negative Ranks	7[a]	5.71	40.00
	Positive Ranks	3[b]	5.00	15.00
	Ties	0[c]		
	Total	10		

a. with_D < without_D
b. with_D > without_D
c. with_D = without_D

Test Statistics[b]

	with_D - without_D
Z	-1.278[a]
Asymp. Sig. (2-tailed)	.201

a. Based on positive ranks.
b. Wilcoxon Signed Ranks Test

The results from the Wilcoxon signed ranks test ($T = 15.0, n = 10, p > 0.05$) indicated that the two testing conditions were not significantly different. Therefore, based on this study, the use of bilingual dictionaries on a math test did not significantly improve scores among limited English proficient students.

2. The results from the analysis are displayed in the SPSS Outputs.

Ranks

		N	Mean Rank	Sum of Ranks
companion - alone	Negative Ranks	8[a]	5.50	44.00
	Positive Ranks	2[b]	5.50	11.00
	Ties	0[c]		
	Total	10		

a. companion < alone
b. companion > alone
c. companion = alone

Test Statistics[b]

	companion - alone
Z	-1.684[a]
Asymp. Sig. (2-tailed)	.092

a. Based on positive ranks.
b. Wilcoxon Signed Ranks Test

The results from the Wilcoxon signed ranks test ($T = 11.0, n = 10, p > 0.05$) indicated that the two conditions were not significantly different. Therefore, based on this study, the presence of a companion in the woods at night did not significantly influence the males' pulse rates.

3. The results from the analysis are displayed in the SPSS Outputs below.

Ranks

		N	Mean Rank	Sum of Ranks
Treatment2 - Treatment1	Negative Ranks	2[a]	5.00	10.00
	Positive Ranks	6[b]	4.33	26.00
	Ties	0[c]		
	Total	8		

a. Treatment2 < Treatment1
b. Treatment2 > Treatment1
c. Treatment2 = Treatment1

Test Statistics[b]

	Treatment2 - Treatment1
Z	-1.123[a]
Asymp. Sig. (2-tailed)	.261

a. Based on negative ranks.
b. Wilcoxon Signed Ranks Test

The results from the Wilcoxon signed ranks test ($T = 10.0, n = 8, p > 0.05$) indicated that the two treatments were not significantly different. Therefore, based on this study, neither treatment program resulted in a significantly higher weight loss among obese female teenagers.

4. The results from the analysis are as follows:

$$T = 50$$
$$x_r = 105 \quad \text{and} \quad s_r = 26.79$$
$$z^* = -2.05$$
$$ES = 0.46$$

This is a reasonably high effect size, which indicates a strong measure of association.

5. For our example, $n = 10$ and $p = 0.05/2$. Thus, $T = 8$ and $K = 9$. The ninth value from the bottom is -1.0 and the ninth value from the top is 7.0. Based on these findings, it is estimated with 95% confidence that the difference in students' number of activities before and after the new program lies between -1.0 and 7.0.

4

COMPARING TWO UNRELATED SAMPLES: THE MANN–WHITNEY U-TEST

4.1 OBJECTIVES

In this chapter, you will learn the following items.

- How to compute the Mann–Whitney U-test.
- How to perform the Mann–Whitney U-test using SPSS.
- How to construct a median confidence interval based on the difference between two independent samples.

4.2 INTRODUCTION

Suppose a teacher wants to know if his first-period's early class time has been reducing student performance. To test his idea, he compares the final exam scores of students in his first-period class with those in his fourth-period class. In this example, each score from one class period is independent of, or unrelated to, the other class period.

The Mann–Whitney U-test is a nonparametric statistical procedure for comparing two samples that are independent, or not related. The Wilcoxon rank sum test is a similar nonparametric test to the Mann–Whitney U-test. The parametric equivalent to these tests is the t-test for independent samples.

In this chapter, we will describe how to perform and interpret a Mann–Whitney U-test for both small samples and large samples. We will also explain how to perform

---

Nonparametric Statistics for Non-Statisticians, Gregory W. Corder and Dale I. Foreman
Copyright © 2009 John Wiley & Sons, Inc.

the procedure using SPSS. Finally, we offer varied examples of these nonparametric statistics from the literature.

4.3 COMPUTING THE MANN–WHITNEY U-TEST STATISTIC

The Mann–Whitney U-test is used to compare two unrelated, or independent, samples. The two samples are combined and rank ordered together. The strategy is to determine if the values from the two samples are randomly mixed in the rank ordering or if they are clustered at opposite ends when combined. A random rank order would mean that the two samples are not different, while a cluster of one sample's values would indicate a difference between them. In Figure 4.1, two sample comparisons illustrate this concept.

The scores in Comparison 1 are rank-ordered in clusters at opposite ends. This suggests that treatment X might be higher than treatment O.	Comparison 1 X X X O X X X X O O O O 1 2 3 4 5 6 7 8 9 10 11 12
The scores in Comparison 2 are spread along the entire distribution. This suggests that there is no clear difference between treatments.	Comparison 2 X O O X X O X O X O X X 1 2 3 4 5 6 7 8 9 10 11 12

FIGURE 4.1

Use Formula 4.1 to determine a Mann–Whitney U-test statistic for each of the two samples. The smaller of the two U statistics is the obtained value.

$$U_i = n_1 n_2 + \frac{n_i(n_i + 1)}{2} - \sum R_i \tag{4.1}$$

where U_i is the test statistic for the sample of interest, n_i is the number of values from the sample of interest, n_1 is the number of values from the first sample, n_2 is the number of values from the second sample, and $\sum R_i$ is the sum of the ranks from the sample of interest.

After the U statistic is computed, it must be examined for significance. We may use a table of critical values (see Table B.4). However, if the number of values in each

COMPUTING THE MANN–WHITNEY U-TEST STATISTIC

sample, n_i, exceeds those available from the table, then a large sample approximation may be performed. For large samples, compute a z-score and use a table with the normal distribution (see Table B.1) to obtain a critical region of z-scores. Formulas 4.2–4.4 are used to find the z-score of a Mann–Whitney U-test for large samples.

$$\bar{x}_U = \frac{n_1 n_2}{2} \quad (4.2)$$

where $\bar{x}_U$ is the mean, n_1 is the number of values from the first sample, and n_2 is the number of values from the second sample.

$$s_U = \sqrt{\frac{n_1 n_2 (n_1 + n_2 + 1)}{12}} \quad (4.3)$$

where s_U is the standard deviation.

$$z^* = \frac{U_i - \bar{x}_U}{s_U} \quad (4.4)$$

where z^* is the z-score for a normal approximation of the data and U_i is the U statistic from the sample of interest.

At this point, the analysis is limited to identifying the presence or absence of a significant difference between the groups and does not describe the strength of the treatment. We can consider the effect size (ES) to determine the degree of association between the groups. We use Formula 4.5 to calculate the effect size.

$$ES = \frac{|z|}{\sqrt{n}} \quad (4.5)$$

where $|z|$ is the absolute value of the z-score and n is the total number of observations.

The effect size ranges from 0 to 1. Cohen (1988) defined the conventions for effect size as small $= 0.10$, medium $= 0.30$, and large $= 0.50$. (Correlation coefficient and effect size are both measures of association. See Chapter 7 concerning correlation for more information on Cohen's assignment of effect size's relative strength.)

4.3.1 Sample Mann–Whitney U-Test (Small Data Samples)

The following data were collected from a study comparing two methods being used to teach reading recovery in the fourth grade. Method 1 was a pullout program in which the children were taken out of the classroom for 30 min. a day, 4 days a week. Method 2 was a small group program in which children were taught in groups of four or five for 45 min. a day in the classroom, 4 days a week. The students were tested using a reading comprehension test after 4 weeks of the program. The test results are shown in Table 4.1.

1. *State the null and research hypotheses.*
    The null hypothesis, shown below, states that there is no tendency of the ranks of one method to be systematically higher or lower than those of the other.

TABLE 4.1

Method 1	Method 2
48	14
40	18
39	20
50	10
41	12
38	102
53	17

The hypothesis is stated in terms of comparison of distributions, not means. The research hypothesis, shown below, states that the ranks of one method are systematically higher or lower than those of the other. Our research hypothesis is a two-tailed, nondirectional hypothesis because it indicates a difference, but in no particular direction.

The null hypothesis is

H_0: There is no tendency for ranks of one method to be significantly higher (or lower) than those of the other.

The research hypothesis is

H_A: The ranks of one method are systematically higher (or lower) than those of the other.

2. *Set the level of risk (or the level of significance) associated with the null hypothesis.*

    The level of risk, also called an alpha (α), is frequently set at 0.05. We will use an alpha of 0.05 in our example. In other words, there is a 95% chance that any observed statistical difference will be real and not due to chance.

3. *Choose the appropriate test statistic.*

    The data are obtained from two independent, or unrelated, samples of fourth-grade children being taught reading. Both the small sample sizes and an existing outlier in the second sample violate our assumptions of normality. Since we are comparing two unrelated, or independent, samples, we will use the Mann–Whitney U-test.

4. *Compute the test statistic.*

    First, combine and rank both data samples together (see Table 4.2). Next, compute the sum of ranks for each method. Method 1 is $\sum R_1$ and Method 2 is $\sum R_2$. Using Table 4.2,

$$\sum R_1 = 7 + 8 + 9 + 10 + 11 + 12 + 13$$
$$= 70$$

TABLE 4.2

Rank	Ordered Scores Score	Sample
1	10	Method 2
2	12	Method 2
3	14	Method 2
4	17	Method 2
5	18	Method 2
6	20	Method 2
7	38	Method 1
8	39	Method 1
9	40	Method 1
10	41	Method 1
11	48	Method 1
12	50	Method 1
13	53	Method 1
14	102	Method 2

and

$$\sum R_2 = 1+2+3+4+5+6+14$$
$$= 35$$

Now, compute the U value for each sample. For sample 1,

$$U_1 = n_1 n_2 + \frac{n_1(n_1+1)}{2} - \sum R_1$$
$$= 7(7) + \frac{7(7+1)}{2} - 70 = 49 + 28 - 70$$
$$= 7$$

and for sample 2,

$$U_2 = n_1 n_2 + \frac{n_2(n_2+1)}{2} - \sum R_2$$
$$= 7(7) + \frac{7(7+1)}{2} - 35 = 49 + 28 - 35$$
$$= 42$$

The Mann–Whitney U-test statistic is the smaller of U_1 and U_2. Therefore, $U = 7$.

5. *Determine the value needed for rejection of the null hypothesis using the appropriate table of critical values for the particular statistic.*

Since the sample sizes are small ($n < 20$), we use Table B.4, which lists the critical values for the Mann–Whitney U. The critical values are found in the table at the point for $n_1 = 7$ and $n_2 = 7$. We set $\alpha = 0.05$. The critical value for the Mann–Whitney U is 8. A calculated value that is less than or equal to 8 will lead us to reject our null hypothesis.

6. *Compare the obtained value to the critical value.*

   The critical value for rejecting the null hypothesis is 8 and the obtained value is $U = 7$. If the critical value equals or exceeds the obtained value, we must reject the null hypothesis. If instead, the critical value is less than the obtained value, we must not reject the null hypothesis. Since the critical value exceeds the obtained value, we must reject the null hypothesis.

7. *Interpret the results.*

   We rejected the null hypothesis, suggesting that a real difference exists between the two methods. In addition, since the sum of the ranks for Method 1 (ΣR_1) was larger than that for Method 2 (ΣR_2), we see that Method 1 had significantly higher scores.

8. *Reporting the results.*

   The reporting of results for the Mann–Whitney U-test should include such information as the sample sizes for each group, the U statistic, the p-value's relation to α, and the sums of ranks for each group.

   For this example, two methods were used to provide students with reading instruction. Method 1 involved a pullout program and Method 2 involved a small group program. Using the ranked reading comprehension test scores, the results indicated a significant difference between the two methods ($U = 7$, $n_1 = 7$, $n_2 = 7$, $p < 0.05$). The sum of ranks for Method 1 ($\Sigma R_1 = 70$) was larger than the sum of ranks for Method 2 ($\Sigma R_2 = 35$). Therefore, we can state that the data support the pullout program as a more effective reading program for teaching comprehension to fourth-grade children at this school.

4.3.2 Performing the Mann–Whitney U-Test Using SPSS

We will analyze the data from the above example using SPSS.

1. *Define your variables.*

   First, click the "Variable View" tab at the bottom of your screen. Then, type the names of your variables in the "Name" column. Unlike the Wilcoxon signed ranks test described in Chapter 2, you cannot simply enter each sample into a separate column to execute the Mann–Whitney U-test. You must use a grouping variable. As shown in Figure 4.2, the first variable is the grouping variable that we called "Method". The second variable that we called "Score" will have our actual values.

COMPUTING THE MANN–WHITNEY U-TEST STATISTIC

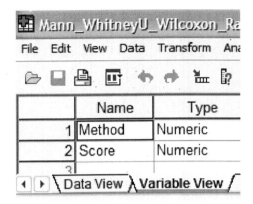

FIGURE 4.2

When establishing a grouping variable, it is often easiest to assign each group a whole number value. In our example, our groups are "Method 1" and "Method 2". Therefore, we must set our grouping variables for the variable "Method". First, we selected the "Values" column and clicked the gray square, as shown in Figure 4.3. Then, we set a value of 1 to equal "Method 1". Now, as soon as we click the "Add" button, we will have set "Method 2" equal to 2 based on the values we inserted above.

FIGURE 4.3

2. *Type in your values.*

Click the "Data View" tab at the bottom of your screen as shown in Figure 4.4. Type in the values for both sets of data in the "Score" column. As you do so, type in the corresponding grouping variable in the "Method" column. For example, all of the values for "Method 2" are signified by a value of 2 in the grouping variable column that we called "Method".

	Method	Score
1	1.00	38.00
2	1.00	39.00
3	1.00	40.00
4	1.00	41.00
5	1.00	48.00
6	1.00	50.00
7	1.00	53.00
8	2.00	10.00
9	2.00	12.00
10	2.00	14.00
11	2.00	17.00
12	2.00	18.00
13	2.00	20.00
14	2.00	102.00
15		

Data View / Variable View

FIGURE 4.4

3. *Analyze your data.*

   As shown in Figure 4.5, use the pull-down menus to choose "Analyze", "Nonparametric Tests", and "2 Independent Samples...".

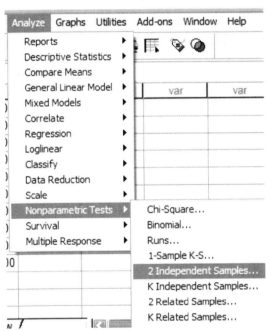

FIGURE 4.5

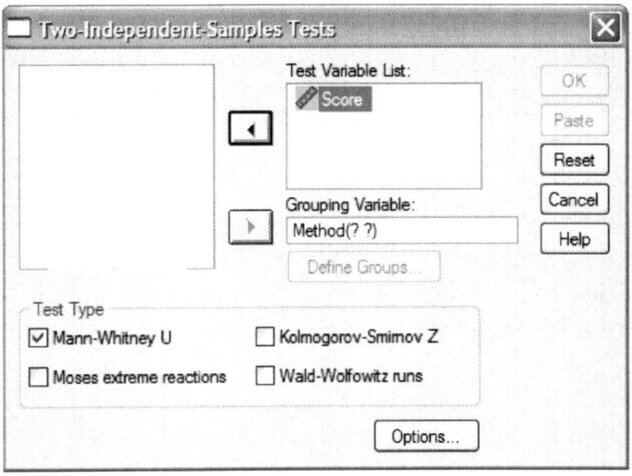

FIGURE 4.6

Use the top arrow button to place your variable with your data values, or dependent variable (DV), in the box labeled "Test Variable List:". Then, use the lower arrow button to place your grouping variable, or independent variable (IV), in the box labeled "Grouping Variable". As shown in Figure 4.6, we have placed the "Score" variable in the "Test Variable List" and the "Method" variable in the "Grouping Variable" box. Click on the "Define Groups..." button to assign a reference value to your independent variable (i.e., "Grouping Variable").

As shown in Figure 4.7, type 1 into the box next to "Group 1" and 2 in the box next to "Group 2". Then, click "Continue". This step references the value labels you created when you defined your grouping variable in step 1. Now that the groups have been assigned, click "OK" to perform the analysis.

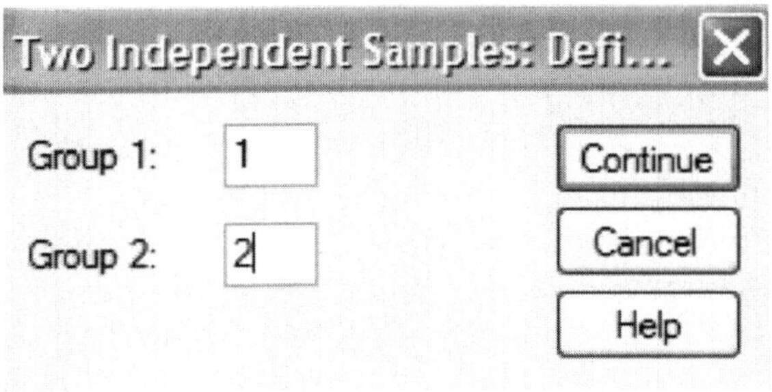

FIGURE 4.7

4. *Interpret the results from the SPSS Output window.*

Ranks

	Method	N	Mean Rank	Sum of Ranks
Score	Method 1	7	10.00	70.00
	Method 2	7	5.00	35.00
	Total	14		

The first SPSS output table provides the sum of ranks and sample sizes for comparing the two groups.

Test Statistics[b]

	Score
Mann-Whitney U	7.000
Wilcoxon W	35.000
Z	-2.236
Asymp. Sig. (2-tailed)	.025
Exact Sig. [2*(1-tailed Sig.)]	.026[a]

a. Not corrected for ties.
b. Grouping Variable: Method

The second SPSS output table provides the Mann–Whitney U-test statistic ($U=7.0$). As described in Figure 4.2, it also returns a similar nonparametric statistic called the Wilcoxon W-test statistic ($W=35.0$). Notice that the Wilcoxon W is the smaller of the two rank sums in the above table.

SPSS returns the critical z-score for large samples. In addition, SPSS calculates the two-tailed significance using two methods. The asymptotic significance is more appropriate with large samples. However, the exact significance is more appropriate with small samples or very poorly distributed data.

Based on the results from SPSS, the ranked reading comprehension test scores of the two methods were significantly different ($U=7$, $n_1=7$, $n_2=7$, $p<0.05$). The sum of ranks for Method 1 ($\sum R_1 = 70$) was larger than the sum of ranks for Method 2 ($\sum R_2 = 35$).

Note: Using the above procedure, SPSS always returns the two-tailed significance. If you were to need a one-tailed significance, simply divide by 2. The one-tailed exact significance in the above example is $0.026/2 = 0.013$.

4.3.3 Confidence Interval for the Difference Between Two Location Parameters

The American Psychological Association (2001) has suggested that researchers report the *confidence interval* for research data. A confidence interval is an inference to a population in terms of an estimation of sampling error. More specifically, it provides a range of values that fall within the population with a level of confidence of $100(1-\alpha)\%$.

COMPUTING THE MANN–WHITNEY U-TEST STATISTIC

A median confidence interval can be constructed based on the difference between two independent samples. It consists of possible values of differences for which we do not reject the null hypothesis at a defined significance level of α.

The test depends on the following assumptions:

1. Data consist of two independent random samples: $X_1, X_2, \ldots, X_n$ from one population and $Y_1, Y_2, \ldots, Y_n$ from the second population.
2. The distribution functions of the two populations are identical except for possible location parameters.

To perform the analysis, set up a table that identifies all possible differences for each possible sample pair such that $D_{ij} = X_i - Y_j$ for (X_i, Y_j). Placing the values for X from smallest to largest across the top and the values for Y from smallest to largest down the side will eliminate the need to order the values of D_{ij} later.

The sample procedure presented below is based on the data from Table 4.2 (small data sample Mann–Whitney U-test) near the beginning of this chapter.

The values from Table 4.2 are arranged in Table 4.3 so that the Method 1 (X) scores are placed in order across the top and the Method 2 (Y) scores are placed in order down the side. Then, the $n_1 n_2$ differences are calculated by subtracting each Y value from each X value. The differences are shown in Table 4.3. Notice that the values of D_{ij} are ordered in the table from highest to lowest starting at the top right and ending at the bottom left.

TABLE 4.3

Y_j				X_i			
	38	39	40	41	48	50	53
10	28	29	30	31	38	40	43
12	26	27	28	29	36	38	41
14	24	25	26	27	34	36	39
17	21	22	23	24	31	33	36
18	20	21	22	23	30	32	35
20	18	19	20	21	28	30	33
102	−64	−63	−62	−61	−54	−52	−49

We use Table B.4 to find the lower limit of the confidence interval, L, and the upper limit, U. For a two-tailed test, L is the $w_{\alpha/2}$th smallest difference and U is the $w_{\alpha/2}$th largest difference that correspond to $\alpha/2$ for n_1 and n_2 for a confidence interval of $(1 - \alpha)$.

For our example, $n_1 = 7$ and $n_2 = 7$. For $0.05/2 = 0.025$, Table B.4 returns $w_{\alpha/2} = 9$. This means that the ninth values from the top and bottom mark the limits of the 95% confidence interval on both ends. Therefore, $L = 19$ and $U = 36$. Based on these results, we are 95% certain that the difference in population median is between 19 and 36.

4.3.4 Sample Mann–Whitney U-Test (Large Data Samples)

The previous comparison of teaching methods for reading recovery was repeated with fifth-grade students. The fifth grade used the same two methods. Method 1 was a pullout program in which the children were taken out of the classroom for 30 min. a day, 4 days a week. Method 2 was a small group program in which children were taught in groups of four or five for 45 min. a day in the classroom, 4 days a week. The students were tested using the same reading comprehension test after 4 weeks of the program. The test results are shown in Table 4.4.

TABLE 4.4

Method 1	Method 2
48	14
40	18
39	20
50	10
41	12
38	102
71	21
30	19
15	100
33	23
47	16
51	82
60	13
59	25
58	24
42	97
11	28
46	9
36	34
27	52
93	70
72	22
57	26
45	8
53	17

1. *State the null and research hypotheses.*

    The null hypothesis, shown below, states that there is no tendency of the ranks of one method to be systematically higher or lower than those of the other. The hypothesis is stated in terms of comparison of distributions, not means. The research hypothesis, shown below, states that the ranks of one method are systematically higher or lower than those of the other. Our research hypothesis

COMPUTING THE MANN–WHITNEY U-TEST STATISTIC

is a two-tailed, nondirectional hypothesis because it indicates a difference, but in no particular direction.

The null hypothesis is

H_0: There is no tendency for ranks of one method to be significantly higher (or lower) than those of the other.

The research hypothesis is

H_A: The ranks of one method are systematically higher (or lower) than those of the other.

2. *Set the level of risk (or the level of significance) associated with the null hypothesis.*

    The level of risk, also called an alpha (α), is frequently set at 0.05. We will use an alpha of 0.05 in our example. In other words, there is a 95% chance that any observed statistical difference will be real and not due to chance.

3. *Choose the appropriate test statistic.*

    The data are obtained from two independent, or unrelated, samples of fifth-grade children being taught reading. Since we are comparing two unrelated, or independent, samples, we will use the Mann–Whitney U-test.

4. *Compute the test statistic.*

    First, combine and rank both data samples together (see Table 4.5). Next, compute the sum of ranks for each method. Method 1 is $\sum R_1$ and Method 2 is $\sum R_2$. Using Table 4.5,

$$\sum R_1 = 779$$

and

$$\sum R_2 = 496$$

Now, compute the U value for each sample. For sample 1,

$$U_1 = n_1 n_2 + \frac{n_1(n_1+1)}{2} - \sum R_1$$
$$= 25(25) + \frac{25(25+1)}{2} - 779 = 625 + 325 - 779$$
$$= 171$$

and for sample 2,

$$U_2 = n_1 n_2 + \frac{n_2(n_2+1)}{2} - \sum R_2$$
$$= 25(25) + \frac{25(25+1)}{2} - 496 = 625 + 325 - 496$$
$$= 454$$

The Mann–Whitney U-test statistic is the smaller of U_1 and U_2. Therefore, $U = 171$.

TABLE 4.5

	Ordered Scores	
Rank	Score	Sample
1	8	Method 2
2	9	Method 2
3	10	Method 2
4	11	Method 1
5	12	Method 2
6	13	Method 2
7	14	Method 2
8	15	Method 1
9	16	Method 2
10	17	Method 2
11	18	Method 2
12	19	Method 2
13	20	Method 2
14	21	Method 2
15	22	Method 2
16	23	Method 2
17	24	Method 2
18	25	Method 2
19	26	Method 2
20	27	Method 1
21	28	Method 2
22	30	Method 1
23	33	Method 1
24	34	Method 2
25	36	Method 1
26	38	Method 1
27	39	Method 1
28	40	Method 1
29	41	Method 1
30	42	Method 1
31	45	Method 1
32	46	Method 1
33	47	Method 1
34	48	Method 1
35	50	Method 1
36	51	Method 1
37	52	Method 2
38	53	Method 1
39	57	Method 1
40	58	Method 1
41	59	Method 1
42	60	Method 1
43	70	Method 2
44	71	Method 1

TABLE 4.5 (Continued)

Rank	Ordered Scores Score	Sample
45	72	Method 1
46	82	Method 2
47	93	Method 1
48	97	Method 2
49	100	Method 2
50	102	Method 2

Since our sample sizes are large, we will approximate them to a normal distribution. Therefore, we will find a z-score for our data using a normal approximation. We must find the mean, $\bar{x}_U$, and the standard deviation, s_U, for the data.

$$\bar{x}_U = \frac{n_1 n_2}{2} = \frac{(25)(25)}{2} = 312.5$$

and

$$s_U = \sqrt{\frac{n_1 n_2 (n_1 + n_2 + 1)}{12}} = \sqrt{\frac{(25)(25)(25 + 25 + 1)}{12}} = \sqrt{\frac{31875}{12}} = 51.54$$

Next, we use the mean, standard deviation, and the U-test statistic to calculate a z-score. Remember, we are testing the hypothesis that there is no difference in the ranks of the scores for two different methods of reading instruction for fifth-grade students.

$$z^* = \frac{U_i - \bar{x}_U}{s_U} = \frac{171 - 312.5}{51.54} = -2.75$$

5. *Determine the value needed for rejection of the null hypothesis using the appropriate table of critical values for the particular statistic.*

   Table B.1 is used to establish the critical region of z-scores. For a two-tailed test with $\alpha = 0.05$, we must not reject the null hypothesis if $-1.96 \leq z^* \leq 1.96$.

6. *Compare the obtained value to the critical value.*

   We find that z^* is not within the critical region of the distribution, $-2.75 < -1.96$. Therefore, we reject the null hypothesis. This suggests a difference between Method 1 and Method 2.

7. *Interpret the results.*
   We rejected the null hypothesis, suggesting that a real difference exists between the two methods. In addition, since the sum of the ranks for Method 1 ($\sum R_1$) was larger than that for Method 2 ($\sum R_2$), we see that Method 1 had significantly higher scores.

   At this point, the analysis is limited to identifying the presence or absence of a significant difference between the groups. In other words, the statistical test's level of significance does not describe the strength of the treatment. The American Psychological Association (2001), however, has called for a measure of the strength called the *effect size*.

   We can consider the effect size for this large sample test to determine the degree of association between the groups. We can use Formula 4.5 to calculate the effect size. For the example, $z = -2.75$ and $n = 50$.

$$ES = \frac{|z|}{\sqrt{n}} = \frac{|-2.75|}{\sqrt{50}}$$
$$= 0.39$$

   Our effect size for the sample difference is 0.39. This value indicates a medium–high level of association between the teaching methods for the reading recovery program with fifth graders.

8. *Reporting the results.*
   For this example, two methods were used to provide fifth-grade students with reading instruction. Method 1 involved a pullout program and Method 2 involved a small group program. Using the ranked reading comprehension test scores, the results indicated a significant difference between the two methods ($U = 171$, $n_1 = 25$, $n_2 = 25$, $p < 0.05$). The sum of ranks for Method 1 ($\sum R_1 = 779$) was larger than the sum of ranks for Method 2 ($\sum R_2 = 496$). Moreover, the effect size for the sample difference was 0.39. Therefore, we can state that the data support the pullout program as a more effective reading program for teaching comprehension to fifth-grade children at this school.

4.4 EXAMPLES FROM THE LITERATURE

Below are varied examples of the nonparametric procedures described in this chapter. We have summarized each study's research problem and researchers' rationale(s) for choosing a nonparametric approach. We encourage you to obtain these studies if you are interested in their results.

- Odaci, H. (2007). Depression, submissive behaviors and negative automatic thoughts in obese Turkish adolescents. *Social Behavior & Personality: An International Journal, 35*(3), 409–416.

Odaci investigated depression, submissive social behaviors, and frequency of automatic negative thoughts in Turkish adolescents. Obese participants were compared with participants of normal weight. After the Shapiro–Wilk statistic revealed that the data were not normally distributed, Odaci applied a Mann–Whitney U-test to compare the groups.

- Bryant, B. K., & Trockel, J. F. (1976). Personal history of psychological stress related to locus of control orientation among college women. *Journal of Consulting and Clinical Psychology, 44*(2), 266–271.

  Bryant and Trockel investigated the impact of stressful life events on undergraduate females' locus of control. The authors compared accrued life changing units for participants with internal control against external using the Mann–Whitney U-test. This nonparametric procedure was selected since the data pertaining to stressful life events were ordinal in nature.

- Re, A. M., Pedron, M., & Cornoldi, C. (2007). Expressive writing difficulties in children described as exhibiting ADHD symptoms. *Journal of Learning Disabilities, 40*(3), 244–255.

  Re, Pedron, and Cornoldi investigated the expressive writing of children with attention-deficit/hyperactivity disorder (ADHD). The authors used a Mann–Whitney U-test to compare students showing symptoms of ADHD behaviors with a control group of students not displaying such behaviors. After examining their data with a Kolmogorov–Smirnov test, the researchers chose the nonparametric procedure due to significant deviations in the data distributions.

- Limb, G. E., & Organista, K. C. (2003). Comparisons between Caucasian students, students of color, and American Indian students on their views on social work's traditional mission, career motivations, and practice preferences. *Journal of Social Work Education, 39*(1), 91–109.

  In an effort to understand the factors that have motivated minority students to enter the social worker profession, Limb and Organista studied data from nearly 7000 students in California entering a social worker program. The authors used a Wilcoxon rank sum test to compare sums of student group ranks. They chose this nonparametric test due to a concern that statistical assumptions were violated regarding sample normality and homogeneity of variances.

- Schulze, E., & Tomal, A. (2006). The chilly classroom: Beyond gender. *College Teaching, 54*(3), 263–270.

  Schulze and Tomal examined classroom climate perceptions among undergraduate students. Since the student questionnaires used an interval scale, they analyzed their findings with a Mann–Whitney U-test.

- Hegedus, K. S. (1999). Providers' and consumers' perspective of nurses' caring behaviours. *Journal of Advanced Nursing, 30*(5), 1090–1096.

  Hegedus performed a pilot study to evaluate a scale designed to examine the caring behaviors of nurses. Care providers were compared with the consumers. She used a Wilcoxon rank sum test in her analysis because study participants were directed to rank the items on the scale.

4.5 SUMMARY

Two samples that are not related may be compared using a nonparametric procedure called the Mann–Whitney U-test (or the Wilcoxon rank sum test). The parametric equivalent to this test is known as the t-test for independent samples.

In this chapter, we described how to perform and interpret a Mann–Whitney U-test for both small samples and large samples. We also explained how to perform the procedure using SPSS. Finally, we offered varied examples of these nonparametric statistics from the literature. The next chapter will involve comparing more than two samples that are related.

4.6 PRACTICE QUESTIONS

1. The data in Table 4.6 were obtained from a reading-level test for first-grade children. Compare the performance gains of the two different methods for teaching reading.

TABLE 4.6

Method	Gain Score	Method	Gain Score
One-on-one	16	Small group	11
One-on-one	13	Small group	2
One-on-one	16	Small group	10
One-on-one	16	Small group	4
One-on-one	13	Small group	9
One-on-one	9	Small group	8
One-on-one	12	Small group	5
One-on-one	12	Small group	6
One-on-one	20	Small group	4
One-on-one	17	Small group	16

Use a two-tailed Mann–Whitney U-test with $\alpha = 0.05$ to determine which method was better for teaching reading. Report your findings.

2. A research study was conducted to see if an active involvement in a hobby had a positive effect on the health of a person who retires after age 65. The data in Table 4.7 describe the health (number of doctor visits in 1 year) for participants who are involved in a hobby almost daily and those who are not.

   Use a one-tailed Mann–Whitney U-test with $\alpha = 0.05$ to determine whether the hobby tends to reduce the need for doctor visits. Report your findings.

TABLE 4.7

No Hobby Group	Hobby Group
12	9
15	5
8	10
11	3
9	4
17	2
5	

3. Table 4.8 shows assessment scores of two different classes who are being taught computer skills using two different methods.

TABLE 4.8

Method 1	Method 2
53	91
41	18
17	14
45	21
44	23
12	99
49	16
50	10

Use a two-tailed Mann–Whitney U-test with $\alpha = 0.05$ to determine which method was better for teaching computer skills. Report your findings.

4. Two methods of teaching reading were compared. Method 1 used the computer to interact with the student and diagnose and remediate the student based upon misconceptions. Method 2 was taught using workbooks in classroom groups. Table 4.9 shows the data obtained on an assessment after 6 weeks of instruction. Calculate the effect size using the z-score from the analysis.

5. Two methods were used to provide instruction in science for the seventh grade. Method 1 included a lab each week and Method 2 had only classroom work with lecture and worksheets. Table 4.10 shows end-of-course test performance for the two methods. Construct a 95% median confidence interval based on the difference between two independent samples to compare the two methods.

TABLE 4.9

Method 1	Method 2
27	9
38	42
15	21
85	83
36	110
95	19
93	29
57	40
63	30
81	23
65	18
77	32
59	101
89	7
41	50
26	37
102	22
55	71
46	16
82	45
24	35
87	28
66	91
12	86
90	20

TABLE 4.10

Method 1	Method 2
15	8
23	15
9	10
12	13
18	17
22	5
17	18
20	7

4.7 SOLUTIONS TO PRACTICE QUESTIONS

1. The results from the analysis are displayed in the SPSS Outputs.

Ranks

	Method	N	Mean Rank	Sum of Ranks
Score	One-on-One	10	14.60	146.00
	Small Group	10	6.40	64.00
	Total	20		

Test Statistics[b]

	Score
Mann-Whitney U	9.000
Wilcoxon W	64.000
Z	-3.116
Asymp. Sig. (2-tailed)	.002
Exact Sig. [2*(1-tailed Sig.)]	.001[a]

a. Not corrected for ties.

b. Grouping Variable: Method

The results from the Mann–Whitney U-test ($U = 9, n_1 = 10, n_2 = 10, p < 0.05$) indicated that the two methods were significantly different. Moreover, the one-on-one method produced a higher sum of ranks ($\Sigma R_1 = 146$) than the small group method ($\Sigma R_2 = 64$). Therefore, based on this study, first-grade children displayed higher reading levels when taught with a one-on-one method.

2. The results from the analysis are displayed in the SPSS Outputs below.

Ranks

	Group	N	Mean Rank	Sum of Ranks
Score	No Hobby	7	9.14	64.00
	Hobby	6	4.50	27.00
	Total	13		

Test Statistics[b]

	Score
Mann-Whitney U	6.000
Wilcoxon W	27.000
Z	-2.149
Asymp. Sig. (2-tailed)	.032
Exact Sig. [2*(1-tailed Sig.)]	.035[a]

a. Not corrected for ties.

b. Grouping Variable: Group

The results from the Mann–Whitney U-test ($U = 6$, $n_1 = 7$, $n_2 = 6$, $p < 0.05$) indicated that the two samples were significantly different. Moreover, the sample with no hobby produced a higher sum of ranks ($\sum R_1 = 64$) than the sample with a hobby ($\sum R_2 = 27$). Therefore, based on this study, a person who retired after age 65 and exhibited daily involvement in a hobby visited the doctor less frequently.

3. The results from the analysis are displayed in the SPSS Outputs below.

Ranks

	Method	N	Mean Rank	Sum of Ranks
Score	Method 1	8	9.50	76.00
	Method 2	8	7.50	60.00
	Total	16		

Test Statistics[b]

	Score
Mann-Whitney U	24.000
Wilcoxon W	60.000
Z	-.840
Asymp. Sig. (2-tailed)	.401
Exact Sig. [2*(1-tailed Sig.)]	.442[a]

a. Not corrected for ties.
b. Grouping Variable: Method

The results from the Mann–Whitney U-test ($U = 24$, $n_1 = 8$, $n_2 = 8$, $p > 0.05$) indicated that the two samples were not significantly different. Therefore, based on this study, neither method resulted in significantly different assessment scores for computer skills.

4. The results from the analysis are as follows:

$$U_1 = 199 \text{ and } U_2 = 426$$
$$x_u = 312.5$$
$$s_u = 51.54$$
$$z^* = -2.20$$
$$ES = 0.31$$

The effect size is moderate.

5. For our example, $n_1 = 8$ and $n_2 = 8$. For $0.05/2 = 0.025$, $w_{\alpha/2} = 14$. Based on these results, we are 95% certain that the median difference between the two methods is between 0 and 11.

5

COMPARING MORE THAN TWO RELATED SAMPLES: THE FRIEDMAN TEST

5.1 OBJECTIVES

In this chapter, you will learn the following items.

- How to compute the Friedman test.
- How to perform contrasts to compare samples.
- How to perform the Friedman test and associated sample contrasts using SPSS.

5.2 INTRODUCTION

Most public school divisions take pride in the percentage of their graduates admitted to college. A large school division might want to determine if these college admission rates are changing or stagnant. The division could compare the percentages of graduates admitted to college from each of its 10 high schools over the past 5 years. Each year would constitute a group, or sample, of percentages from each school. In other words, the study would include five groups and each group would include 10 values.

The samples in the example are dependent, or related, since each school has a percentage for each year. The Friedman test is a nonparametric statistical procedure

---

Nonparametric Statistics for Non-Statisticians, Gregory W. Corder and Dale I. Foreman
Copyright © 2009 John Wiley & Sons, Inc.

for comparing more than two samples that are related. The parametric equivalent to this test is the repeated measures analysis of variance (ANOVA).

When the Friedman test leads to significant results, at least one of the samples is different from the other samples. However, the test does not identify where the difference(s) occur. Moreover, it does not identify how many differences occur. To identify the particular differences between sample pairs, a researcher might use sample contrasts, or post hoc tests, to analyze the specific sample pairs for significant difference(s). The Wilcoxon signed ranks test (see Chapter 3) is a useful method for performing sample contrasts between individual sample sets.

In this chapter, we will describe how to perform and interpret a Friedman test followed with sample contrasts. We will also explain how to perform the procedures using SPSS. Finally, we offer varied examples of these nonparametric statistics from the literature.

5.3 COMPUTING THE FRIEDMAN TEST STATISTIC

The Friedman test is used to compare more than two dependent samples. When stating our hypotheses, we state them in terms of the population. Moreover, we examine the population medians, θ_i, when performing the Friedman test.

To compute the Friedman test statistic, F_r, we begin by creating a table of our data. List the research subjects to create the rows. Place the values for each condition in columns next to the appropriate subjects. Then, rank the values for each subject across each condition. If there are no ties from the ranks, use Formula 5.1 to determine the Friedman test statistic, F_r.

$$F_r = \left[\frac{12}{nk(k+1)} \sum_{i=1}^{k} R_i^2 \right] - 3n(k+1) \tag{5.1}$$

where n is the number of rows, or subjects; k is the number of columns, or conditions; and R_i is the sum of the ranks from column, or condition, i.

If ranking of values results in any ties, use Formula 5.2 to determine the Friedman test statistic, F_r.

$$F_r = \frac{n(k-1)\left[\sum_{i=1}^{k} \frac{R_i^2}{n} - C_F \right]}{\sum r_{ij}^2 - C_F} \tag{5.2}$$

where n is the number of rows, or subjects; k is the number of columns, or conditions; R_i is the sum of the ranks from column, or condition, i; C_F is the ties correction, $(1/4)nk(k+1)^2$; and r_{ij} is the rank corresponding to subject j in column i.

The degrees of freedom for the Friedman test are determined by using Formula 5.3.

$$df = k-1 \tag{5.3}$$

where df is the degrees of freedom and k is the number of groups.

COMPUTING THE FRIEDMAN TEST STATISTIC

Once the test statistic, F_r, is computed, it can be compared to a table of critical values (see Table B.5) to examine the groups for significant differences. However, if the number of groups, k, or the number of values in a group, n, exceeds those available from the table, then a large sample approximation may be performed. Use a table with the chi-square distribution (see Table B.2) to obtain a critical value when performing a large sample approximation.

If the F_r statistic is not significant, then no differences exist between any of the related conditions. However, if the F_r statistic is significant, then a difference exists between at least two of the conditions. Therefore, a researcher might use sample contrasts between individual pairs of conditions, or post hoc tests, to determine which of the condition pairs are significantly different.

When performing multiple sample contrasts, the Type I error rate tends to become inflated. Therefore, the initial level of risk, or α, must be adjusted. We recommend the Bonferroni procedure, shown in Formula 5.4, to adjust α.

$$\alpha_B = \frac{\alpha}{k} \tag{5.4}$$

where α_B is the adjusted level of risk, α is the original level of risk, and k is the number of comparisons.

5.3.1 Sample Friedman Test (Small Data Samples Without Ties)

A manager is struggling with the chronic tardiness of her seven employees. She tries two strategies to improve employee timeliness. First, over the course of a month, she punishes employees with a $10 paycheck deduction for each day that they arrive late. Second, the following month, she punishes employees by docking their pay $20 for each day that they do not arrive on time.

Table 5.1 shows the number of times each employee was late in a given month. Baseline shows the employees' monthly tardiness before the strategies. Month 1 shows the employees' monthly tardiness after a month of the $10 paycheck deductions. Month 2 shows the employees' monthly tardiness after a month of the $20 paycheck deductions.

TABLE 5.1

Employee	Monthly Tardiness		
	Baseline	Month 1	Month 2
1	16	13	12
2	10	5	2
3	7	8	9
4	13	11	5
5	17	2	6
6	10	7	9
7	11	6	7

We want to determine if either of the paycheck deduction strategies reduced employee tardiness. Since the sample sizes are small ($n < 20$), we require a nonparametric test. The Friedman test is the best statistic to analyze the data and test the hypothesis.

1. *State the null and research hypotheses.*

    The null hypothesis, shown below, states that neither of the manager's strategies will change employee tardiness. The research hypothesis states that one or both of the manager's strategies will reduce employee tardiness.

    The null hypothesis is

    H_0: $\theta_B = \theta_{M1} = \theta_{M2}$

    The research hypothesis is

    H_A: One or both of the manager's strategies will reduce employee tardiness.

2. *Set the level of risk (or the level of significance) associated with the null hypothesis.*

    The level of risk, also called an alpha (α), is frequently set at 0.05. We will use an alpha of 0.05 in our example. In other words, there is a 95% chance that any observed statistical difference will be real and not due to chance.

3. *Choose the appropriate test statistic.*

    The data are obtained from three dependent, or related, conditions that report employees' number of monthly tardiness. The three samples are small with some violations of our assumptions of normality. Since we are comparing three dependent conditions, we will use the Friedman test.

4. *Compute the test statistic.*

    First, rank the values from each employee, or subject (see Table 5.2).

TABLE 5.2

	Ranks of Monthly Tardiness		
Employee	Baseline	Month 1	Month 2
1	3	2	1
2	3	2	1
3	1	2	3
4	3	2	1
5	3	1	2
6	3	1	2
7	3	1	2

Next, compute the sum of ranks for each condition. The ranks in each group are added to obtain a total R value for the group.

For the Baseline condition,

$$R_B = 3 + 3 + 1 + 3 + 3 + 3 + 3 = 19$$

COMPUTING THE FRIEDMAN TEST STATISTIC

For Month 1,
$$R_{M1} = 2+2+2+2+1+1+1 = 11$$

For Month 2,
$$R_{M2} = 1+1+3+1+2+2+2 = 12$$

These R values are used to compute the F_r test statistic. Use Formula 5.1 since there were no ties involved in the ranking.

$$F_r = \left[\frac{12}{nk(k+1)}\sum_{i=1}^{k}R_i^2\right] - 3n(k+1)$$

$$= \left(\frac{12}{7(3)(3+1)}\right)(19^2 + 11^2 + 12^2) - 3(7)(3+1)$$

$$= \left(\frac{12}{84}\right)(361 + 121 + 144) - 84$$

$$= (0.1429)(626) - 84$$

$$= 89.4286 - 84$$

$$= 5.429$$

5. *Determine the value needed for rejection of the null hypothesis using the appropriate table of critical values for the particular statistic.*

   We will use the table of critical values for the Friedman test (see Table B.5) since it includes the number of groups, k, and the number of samples, n, for our data. In this case, we look for the critical value for $k = 3$ and $n = 7$ with $\alpha = 0.05$. Table B.5 returns a critical value for the Friedman test of 7.14.

6. *Compare the obtained value to the critical value.*

   The critical value for rejecting the null hypothesis is 7.14 and the obtained value is $F_r = 5.429$. If the critical value is less than or equal to the obtained value, we must reject the null hypothesis. If instead, the critical value exceeds the obtained value, we do not reject the null hypothesis. Since the critical value exceeds the obtained value, we do not reject the null hypothesis.

7. *Interpret the results.*

   We did not reject the null hypothesis, suggesting that no significant difference exists between any of the three conditions. Therefore, no further comparisons are necessary with these data.

8. *Reporting the results.*

   The reporting of results for the Friedman test should include such information as the number of subjects, the F_r statistic, degrees of freedom, and p-value's relation to α.

   For this example, the frequencies of employees' ($n = 7$) tardiness were compared over three conditions. The Friedman test was not significant

($F_{r(2)} = 5.429$, $p > 0.05$). Therefore, we can state that the data do not support punishing tardy employees with $10 or $20 paycheck deductions.

5.3.2 Sample Friedman Test (Small Data Samples with Ties)

After the manager's failure to reduce employee tardiness with paycheck deductions, she decided to try a different approach. This time, she rewarded employees when they arrived to work on time. Again, she tries two strategies to improve employee timeliness. First, over the course of a month, she rewards employees with a $10 bonus for each day that they arrive on time. Second, the following month, she rewards employees with a $20 bonus for each day that they arrive on time.

Table 5.3 shows the number of times each employee was late in a given month. Baseline shows the employees' monthly tardiness before any of the strategies in either example. Month 1 shows the employees' monthly tardiness after a month of the $10 bonuses. Month 2 shows the employees' monthly tardiness after a month of the $20 bonuses.

TABLE 5.3

Employee	Monthly Tardiness		
	Baseline	Month 1	Month 2
1	16	17	11
2	10	5	2
3	7	8	0
4	13	9	5
5	17	2	2
6	10	10	9
7	11	6	5

We want to determine if either of the strategies reduced employee tardiness. Again, since the sample sizes are small ($n < 20$), we require a nonparametric test. The Friedman test is the best statistic to analyze the data and test the hypothesis.

1. *State the null and research hypotheses.*

    The null hypothesis, shown below, states that neither of the manager's strategies will change employee tardiness. The research hypothesis states that one or both of the manager's strategies will reduce employee tardiness.

    The null hypothesis is

    H_0: $\theta_B = \theta_{M1} = \theta_{M2}$

    The research hypothesis is

    H_A: One or both of the manager's strategies will reduce employee tardiness.

2. *Set the level of risk (or the level of significance) associated with the null hypothesis.*

    The level of risk, also called an alpha (α), is frequently set at 0.05. We will use an alpha of 0.05 in our example. In other words, there is a 95% chance that any observed statistical difference will be real and not due to chance.

COMPUTING THE FRIEDMAN TEST STATISTIC

3. *Choose the appropriate test statistic.*

   The data are obtained from three dependent, or related, conditions that report employees' number of monthly tardiness. The three samples are small with some violations of our assumptions of normality. Since we are comparing three dependent conditions, we will use the Friedman test.

4. *Compute the test statistic.*

   First, rank the values from each employee, or subject (see Table 5.4). Next, compute the sum of ranks for each condition. The ranks in each group are added to obtain a total R value for the group.

TABLE 5.4

	Ranks of Monthly Tardiness		
Employee	Baseline	Month 1	Month 2
1	2	3	1
2	3	2	1
3	2	3	1
4	3	2	1
5	3	1.5	1.5
6	2.5	2.5	1
7	3	2	1

For the Baseline condition,
$$R_B = 2+3+2+3+3+2.5+3 = 18.5$$

For Month 1,
$$R_{M1} = 3+2+3+2+1.5+2.5+2 = 16$$

For Month 2,
$$R_{M2} = 1+1+1+1+1.5+1+1 = 7.5$$

These R values are used to compute the F_r test statistic. Since there were ties involved in the rankings, we must use Formula 5.2. Finding the values for C_F and $\sum r_{ij}^2$ first will simplify the calculation.

$$\begin{aligned} C_F &= (1/4)nk(k+1)^2 \\ &= (1/4)(7)(3)(3+1)^2 \\ &= 84 \end{aligned}$$

To find $\sum r_{ij}^2$, square all of the ranks. Then, add all of the squared ranks together (see Table 5.5).

$$\begin{aligned} \sum r_{ij}^2 &= 50.25 + 38.50 + 8.25 \\ &= 97.0 \end{aligned}$$

COMPARING MORE THAN TWO RELATED SAMPLES: THE FRIEDMAN TEST

TABLE 5.5

Employee	Ranks of Monthly Tardiness		
	Baseline	Month 1	Month 2
1	4	9	1
2	9	4	1
3	4	9	1
4	9	4	1
5	9	2.25	2.25
6	6.25	6.25	1
7	9	4	1
$\sum r_i^2$	50.25	38.50	8.25

Now that we have C_F and $\sum r_{ij}^2$ we are ready for Formula 5.2.

$$F_r = \frac{n(k-1)\left[\sum_{i=1}^{k}\frac{R_i^2}{n} - C_F\right]}{\sum r_{ij}^2 - C_F}$$

$$= \frac{7(3-1)\left[\frac{18.5}{7} + \frac{16.0}{7} + \frac{7.5}{7} - 84\right]}{97-84}$$

$$= \frac{7(2)[48.89 + 36.57 + 8.04 - 84]}{13}$$

$$= \frac{7(2)9.5}{13}$$

$$= 10.23$$

5. *Determine the value needed for rejection of the null hypothesis using the appropriate table of critical values for the particular statistic.*

   We will use the table of critical values for the Friedman test (see Table B.5) since it includes the number of groups, k, and the number of samples, n, for our data. In this case, we look for the critical value for $k = 3$ and $n = 7$ with $\alpha = 0.05$. Table B.5 returns a critical value for the Friedman test of 7.14.

6. *Compare the obtained value to the critical value.*

   The critical value for rejecting the null hypothesis is 7.14 and the obtained value is $F_r = 10.23$. If the critical value is less than or equal to the obtained value, we must reject the null hypothesis. If instead, the critical value exceeds the obtained value, we do not reject the null hypothesis. Since the obtained value exceeds the critical value, we reject the null hypothesis.

7. *Interpret the results.*

   We rejected the null hypothesis, suggesting that a significant difference exists between one or more of the three conditions. In particular, both strategies seemed to result in less tardiness among employees. However, describing specific differences in this manner is speculative. Therefore, we need a technique for statistically identifying difference between conditions, or contrasts.

7b. *Sample contrasts, or post hoc tests.*

The Friedman test identifies if a statistical difference exists; however, it does not identify how many differences exist and which conditions are different. To identify which conditions are different and which are not, we use a procedure called contrasts, or post hoc tests. An appropriate test to use when comparing two related conditions at a time is the Wilcoxon signed ranks test described in Chapter 2.

It is important to note that performing several Wilcoxon signed ranks tests has a tendency to inflate the Type I error rate. In our example, we would compare three groups, $k = 3$. At an $\alpha = 0.05$, the Type I error rate would equal $1 - (1 - 0.05)^3 = 0.14$.

To compensate for this error inflation, we suggest using the Bonferroni procedure (see Formula 5.4). With this technique, we use a corrected α with the Wilcoxon signed ranks tests to determine significant differences between conditions. For our example, we are only comparing Month 1 and Month 2 against the Baseline. We are not comparing Month 1 against Month 2. Therefore, we are only making two comparisons and $k = 2$.

$$\alpha_B = \frac{\alpha}{k} = \frac{0.05}{2} = 0.025$$

When we compare the three samples with the Wilcoxon signed ranks tests and α_B (see Chapter 2), we obtain the results presented in Table 5.6. Notice that the significance is one tailed, or directional, since we were seeking a decline in tardiness.

TABLE 5.6

Condition Comparison	Wilcoxon T Statistic	Rank Sum Difference	One-Tailed Significance
Baseline–Month 1	3.0	18.0 – 3.0 = 15.0	0.057
Baseline–Month 2	0.0	28.0 – 0.0 = 28.0	0.009

Using $\alpha_B = 0.025$, we notice that the Baseline–Month 1 comparison does not demonstrate a significant difference ($p > 0.025$). However, the Baseline–Month 2 comparison does demonstrate a significant difference ($p < 0.025$). Therefore, the data indicate that the $20 bonus reduces tardiness while the $10 bonus does not.

Note: If you are not comparing all of the samples for the Friedman test, then k is only the number of comparisons you are making with the Wilcoxon signed ranks tests. Therefore, comparing fewer samples will increase the chances of finding a significant difference.

8. *Reporting the results.*

The reporting of results for the Friedman test should include such information as the number of subjects, the F_r statistic, degrees of freedom, and *p*-value's relation to α.

For this example, the frequencies of employees' ($n=7$) tardiness were compared over three conditions. The Friedman test was significant ($F_{r(2)} = 10.23$, $p < 0.05$). In addition, follow-up contrasts using Wilcoxon signed ranks tests revealed that $20 bonus reduces tardiness, while the $10 bonus does not.

5.3.3 Performing the Friedman Test Using SPSS

We will analyze the data from the above example using SPSS.

1. *Define your variables.*

    First, click the "Variable View" tab at the bottom of your screen. Then, type the names of your variables in the "Name" column. As shown in Figure 5.1, we have named our variables "Baseline", "Month_1", and "Month_2".

	Name	Type
1	Baseline	Numeric
2	Month_1	Numeric
3	Month_2	Numeric
4		

FIGURE 5.1

2. *Type in your values.*

    Click the "Data View" tab at the bottom of your screen and type your data under the variable names. As shown in Figure 5.2, we are comparing "Baseline", "Month_1", and "Month_2".

3. *Analyze your data.*

    As shown in Figure 5.3, use the pull-down menus to choose "Analyze", "Nonparametric Tests", and "K Related Samples...".

    Select each of the variables that you want to compare and click the button in the middle to move it to the "Test Pair(s) List" as shown in Figure 5.4. Notice that the "Friedman" box is checked by default. After the variables are in the "Test Pair(s) List", click "OK" to perform the analysis.

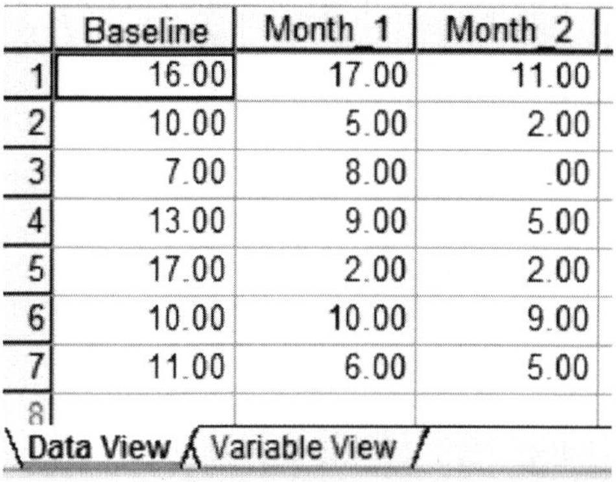

COMPUTING THE FRIEDMAN TEST STATISTIC

	Baseline	Month_1	Month_2
1	16.00	17.00	11.00
2	10.00	5.00	2.00
3	7.00	8.00	.00
4	13.00	9.00	5.00
5	17.00	2.00	2.00
6	10.00	10.00	9.00
7	11.00	6.00	5.00
8			

FIGURE 5.2

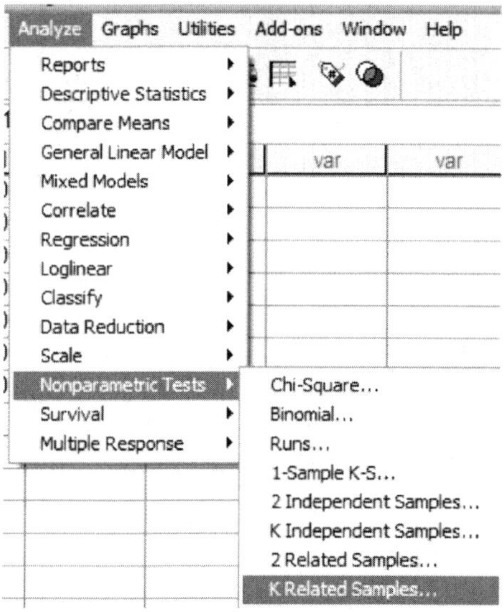

FIGURE 5.3

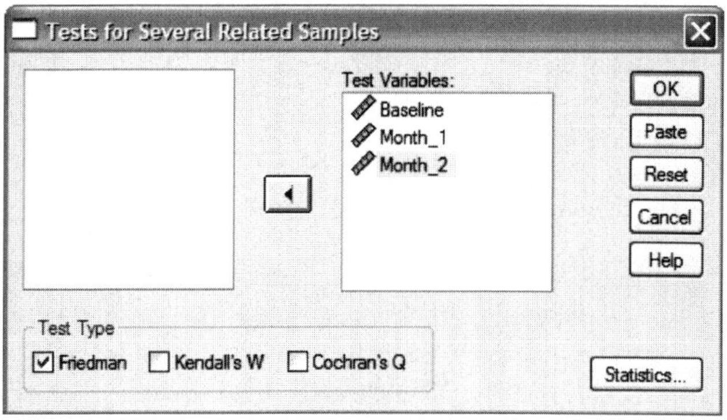

FIGURE 5.4

4. *Interpret the results from the SPSS Output window.*

Ranks

	Mean Rank
Baseline	2.64
Month_1	2.29
Month_2	1.07

Test Statistics[a]

N	7
Chi-Square	10.231
df	2
Asymp. Sig.	.006

a. Friedman Test

The first output table provides the mean ranks of each condition. The second output table provides the Friedman test statistic, 10.231. Since this test uses a chi-square distribution, SPSS calls the F_r statistic "Chi-Square". This table also returns the number of subjects ($n = 7$), degrees of freedom (df = 2), and the significance ($p = 0.006$).

Based on the results from SPSS, three conditions were compared among employees ($n = 7$). The Friedman test was significant ($F_{r(2)} = 10.23, p < 0.05$). To compare individual pairs of conditions, contrasts may be used.

Note: To perform Wilcoxon signed ranks tests for sample contrasts, remember to use your corrected level of risk, α_B, when examining your significance.

5.3.4 Sample Friedman Test (Large Data Samples Without Ties)

After hearing of the manager's success, the head office transferred her to a larger branch office. The transfer was strategic because this larger branch is dealing with tardiness issues among employees. The manager suggests that she use the same successful incentives for employee timeliness. Due to financial limitations, however, she is limited to offering employees smaller bonuses. First, over the course of a month, she rewards employees with a $2 bonus for each day that they arrive on time. Second, the following month, she rewards employees with a $5 bonus for each day that they arrive on time.

Table 5.7 shows the number of times each employee was late in a given month. Baseline shows the employees' monthly tardiness before any of the strategies. Month 1 shows the employees' monthly tardiness after a month of the $2 bonuses. Month 2 shows the employees' monthly tardiness after a month of the $5 bonuses.

COMPUTING THE FRIEDMAN TEST STATISTIC

TABLE 5.7

Employee	Monthly Tardiness		
	Baseline	Month 1	Month 2
1	16	13	12
2	10	5	12
3	7	8	9
4	13	11	5
5	17	2	6
6	10	17	9
7	11	6	7
8	9	8	10
9	12	13	7
10	10	7	8
11	5	8	4
12	11	6	12
13	13	7	6
14	4	6	10
15	10	5	7
16	8	9	6
17	8	3	12
18	15	10	12
19	2	3	11
20	2	4	5
21	10	3	1
22	12	5	6
23	8	12	3
24	11	6	1
25	4	14	5

We want to determine if either of the paycheck bonus strategies reduced employee tardiness. Since the sample sizes are large ($n > 20$), we will use chi-square for the critical value. The Friedman test is the best statistic to analyze the data and test the hypothesis.

1. *State the null and research hypotheses.*

    The null hypothesis, shown below, states that neither of the manager's strategies will change employee tardiness. The research hypothesis states that one or both of the manager's strategies will reduce employee tardiness.

    The null hypothesis is

    $H_0: \theta_B = \theta_{M1} = \theta_{M2}$

    The research hypothesis is

    H_A: One or both of the manager's strategies will reduce employee tardiness.

2. *Set the level of risk (or the level of significance) associated with the null hypothesis.*

The level of risk, also called an alpha (α), is frequently set at 0.05. We will use an alpha of 0.05 in our example. In other words, there is a 95% chance that any observed statistical difference will be real and not due to chance.

3. *Choose the appropriate test statistic.*

The data are obtained from three dependent, or related, conditions that report employees' number of monthly tardiness. Since we are comparing three dependent conditions, we will use the Friedman test.

4. *Compute the test statistic.*

First, rank the values from each employee, or subject (see Table 5.8). Next, compute the sum of ranks for each condition. The ranks in each group are added to obtain a total R value for the group.

For the Baseline condition,
$$R_B = 56$$
For Month 1,
$$R_{M1} = 46$$
For Month 2,
$$R_{M2} = 48$$

TABLE 5.8

Employee	Ranks of Monthly Tardiness		
	Baseline	Month 1	Month 2
1	3	2	1
2	2	1	3
3	1	2	3
4	3	2	1
5	3	1	2
6	2	3	1
7	3	1	2
8	2	1	3
9	2	3	1
10	3	1	2
11	2	3	1
12	2	1	3
13	3	2	1
14	1	2	3
15	3	1	2
16	2	3	1
17	2	1	3
18	3	1	2
19	1	2	3
20	1	2	3
21	3	2	1
22	3	1	2
23	2	3	1
24	3	2	1
25	1	3	2

EXAMPLES FROM THE LITERATURE

These R values are used to compute the F_r test statistic. Use Formula 5.1 since there were no ties involved in the ranking.

$$F_r = \left[\frac{12}{nk(k+1)} \sum_{i=1}^{k} R_i^2\right] - 3n(k+1)$$

$$= \left(\frac{12}{25(3)(3+1)}\right)(56^2 + 46^2 + 48^2) - 3(25)(3+1)$$

$$= \left(\frac{12}{300}\right)(3136 + 2116 + 2304) - 300$$

$$= (0.04)(7556) - 300$$

$$= 302.24 - 300$$

$$= 2.24$$

5. *Determine the value needed for rejection of the null hypothesis using the appropriate table of critical values for the particular statistic.*

 Since the data are a large sample, we will use the chi-square distribution (see Table B.2) to find the critical value for the Friedman test. In this case, we look for the critical value for df = 2 and $\alpha = 0.05$. Using the table, the critical value for rejecting the null hypothesis is 5.99.

6. *Compare the obtained value to the critical value.*

 The critical value for rejecting the null hypothesis is 5.99 and the obtained value is $F_r = 2.24$. If the critical value is less than or equal to the obtained value, we must reject the null hypothesis. If instead, the critical value exceeds the obtained value, we do not reject the null hypothesis. Since the critical value exceeds the obtained value, we do not reject the null hypothesis.

7. *Interpret the results.*

 We did not reject the null hypothesis, suggesting that no significant difference exists between one or more of the three conditions. In particular, the data suggest that neither strategy seemed to result in less tardiness among employees.

8. *Reporting the results.*

 The reporting of results for the Friedman test should include such information as the number of subjects, the F_r statistic, degrees of freedom, and p-value's relation to α.

 For this example, the frequencies of employees' ($n = 25$) tardiness were compared over three conditions. The Friedman test was not significant ($F_{r(2)} = 2.24$, $p > 0.05$). Therefore, we can state that the data do not support providing employees with the \$2 or \$5 paycheck incentive.

5.4 EXAMPLES FROM THE LITERATURE

Below are varied examples of the nonparametric procedures described in this chapter. We have summarized each study's research problem and researchers' rationale(s) for choosing a nonparametric approach. We encourage you to obtain these studies if you are interested in their results.

- Marston, D. (1996). A comparison of inclusion only, pull-out only, and combined service models for students with mild disabilities. *The Journal of Special Education, 30*(2), 121–132.

  Marston examined teachers' attitudes toward three models for servicing elementary students with mild disabilities. He compared special education resource teachers' ratings of the three models (inclusion only, combined services, and pullout only) using a Friedman test. He chose this nonparametric test because the teachers' attitude responses were based on rankings. When the Friedman test produced significant results, he modified the α with the Bonferroni procedure to avoid a ballooned Type I error rate with follow-up comparisons.

- Savignon, S. J., & Sysoyev, P. V. (2002). Sociocultural strategies for a dialogue of cultures. *The Modern Language Journal, 86*(4), 508–524.

  From a Russian high school's English as a foreign language program, Savignon and Sysoyev examined 30 students' responses to explicit training in coping strategies for particular social and cultural situations. Since the researchers considered each student a block in a randomized block study, they used a Friedman test to compare the 30 students, or groups. A nonparametric test was chosen because there were only two possible responses for each strategy (1 = strategy was difficult; 0 = strategy was not difficult). When the Friedman test produced significant results, they used a follow-up sign test to examine each pair for differences in response to find out which of the seven strategies were more difficult than others.

- Cady, J., Meier, S. L., & Lubinski, C. A. (2006). Developing mathematics teachers: The transition from preservice to experienced teacher. *The Journal of Educational Research, 99*(5), 295–305.

  Cady, Meier, and Lubinski examined math teachers' beliefs about the teaching and learning of mathematics over time. Since their sample size was small ($n = 12$), they used a Friedman test to compare scores of participants' survey responses. When participants' scores on the surveys differed significantly, the researchers performed follow-up pairwise analyses with the Wilcoxon signed ranks test.

- Hardré, P. L., Crowson, H. M., Xie, K., & Ly, C. (2007). Testing differential effects of computer-based, web-based and paper-based administration of questionnaire research instruments. *British Journal of Educational Technology, 38*(1), 5–22.

  Hardré, Crowson, Xie, and Ly sought to determine if computer-based, paper-based, and web-based test administrations produce the same results. They compared university students' performance on each of the three test styles. Since normality violations were observed, the researchers used a Friedman test to compare correlations of the three methods. Follow-up contrast tests were not performed since no significant differences were observed.

5.5 SUMMARY

More than two samples that are related may be compared using the Friedman test. The parametric equivalent to this test is known as the repeated measures analysis of variance (ANOVA). When the Friedman test produces significant results, it does not identify which or how many pairs of conditions are significantly different. The Wilcoxon signed ranks test, with a Bonferroni procedure to avoid Type I error rate inflation, is a useful method for comparing individual condition pairs.

In this chapter, we described how to perform and interpret a Friedman test followed with sample contrasts. We also explained how to perform the procedures using SPSS. Finally, we offered varied examples of these nonparametric statistics from the literature. The next chapter will involve a nonparametric procedure for comparing more than two unrelated samples.

5.6 PRACTICE QUESTIONS

1. A graduate student performed a pilot study for his dissertation. He wanted to examine the effects of animal companionship on elderly males. He selected 10 male participants from a nursing home. Then he used an ABAB research design where A represented a week with the absence of a cat and B represented a week with the presence of a cat. At the end of each week, he administered a 20-point survey to measure quality of life satisfaction. The survey results are presented in Table 5.9.

TABLE 5.9

Participant	Week 1	Week 2	Week 3	Week 4
1	7	6	8	9
2	9	8	10	7
3	15	18	16	17
4	7	6	8	9
5	7	8	10	11
6	10	14	13	11
7	12	19	11	13
8	7	4	2	5
9	8	7	9	5
10	12	16	14	15

Use a Friedman test to determine if one or more of the groups are significantly different. Since this is pilot study, use $\alpha = 0.10$. If a significant difference exists, use Wilcoxon signed ranks tests to identify which groups are significantly different. Use the Bonferroni procedure to limit the Type I error rate. Report your findings.

TABLE 5.10

	Number of Curl-Ups in 1 min		
Participant	Baseline	Month 1	Month 2
1	66	67	69
2	49	50	56
3	51	52	49
4	65	65	69
5	42	43	46
6	38	39	40
7	33	31	39
8	41	41	44
9	46	47	48
10	45	46	46
11	36	33	34
12	51	55	67

2. A physical education teacher conducted an action research project to examine a strength and conditioning program. Using 12 male participants, she measures the number of curl-ups they could do in 1 min. She measured their performance before the programs. Then, she measured their performance at 1-month intervals. Table 5.10 presents the performance results.

Use a Friedman test with $\alpha = 0.05$ to determine if one or more of the groups are significantly different. The teacher is expecting performance gains, so if a significant difference exists, use one-tailed Wilcoxon signed ranks tests to identify which groups are significantly different. Use the Bonferroni procedure to limit the Type I error rate. Report your findings.

5.7 SOLUTIONS TO PRACTICE QUESTIONS

1. The results from the Friedman test are displayed in the SPSS Output.

Test Statistics[a]

N	10
Chi-Square	2.160
df	3
Asymp. Sig.	.540

a. Friedman Test

According to the data, the results from the Friedman test indicated that the four conditions were not significantly different ($F_{r(3)} = 2.160$, $p > 0.10$). Therefore, no follow-up contrasts are needed.

SOLUTIONS TO PRACTICE QUESTIONS

2. The results from the Friedman test are displayed in the SPSS Output below.

Test Statistics[a]

N	12
Chi-Square	10.978
df	2
Asymp. Sig.	.004

a. Friedman Test

According to the data, the results from the Friedman test indicated that one or more of the three groups are significantly different ($F_{r(2)} = 10.978$, $p < 0.05$). Therefore, we must examine each set of samples with follow-up contrasts to find the differences between groups. We compare the samples with Wilcoxon signed ranks tests. Since there are $k = 3$ groups, use $\alpha_B = 0.0167$ to avoid Type I error rate inflation. The results from the Wilcoxon signed ranks tests are displayed in the SPSS Outputs below.

Ranks

		N	Mean Rank	Sum of Ranks
Month_1 - Baseline	Negative Ranks	2[a]	8.50	17.00
	Positive Ranks	8[b]	4.75	38.00
	Ties	2[c]		
	Total	12		
Month_2 - Month_1	Negative Ranks	1[d]	6.00	6.00
	Positive Ranks	10[e]	6.00	60.00
	Ties	1[f]		
	Total	12		
Month_2 - Baseline	Negative Ranks	2[g]	3.50	7.00
	Positive Ranks	10[h]	7.10	71.00
	Ties	0[i]		
	Total	12		

a. Month_1 < Baseline
b. Month_1 > Baseline
c. Month_1 = Baseline
d. Month_2 < Month_1
e. Month_2 > Month_1
f. Month_2 = Month_1
g. Month_2 < Baseline
h. Month_2 > Baseline
i. Month_2 = Baseline

Test Statistics[b]

	Month_1 - Baseline	Month_2 - Month_1	Month_2 - Baseline
Z	-1.111[a]	-2.410[a]	-2.522[a]
Asymp. Sig. (2-tailed)	.266	.016	.012

a. Based on negative ranks.
b. Wilcoxon Signed Ranks Test

a. *Baseline–Month 1 comparison.*

The results from the Wilcoxon signed ranks test ($T = 17.0$, $n = 12$, $p > 0.0167$) indicated that the two samples were not significantly different.

b. *Month 1–Month 2 comparison.*

The results from the Wilcoxon signed ranks test ($T = 6.0$, $n = 12$, $p < 0.0167$) indicated that the two samples were significantly different.

c. *Baseline–Month 2 comparison.*

The results from the Wilcoxon signed ranks test ($T = 7.0$, $n = 12$, $p < 0.0167$) indicated that the two samples were significantly different.

6

COMPARING MORE THAN TWO UNRELATED SAMPLES: THE KRUSKAL–WALLIS H-TEST

6.1 OBJECTIVES

In this chapter, you will learn the following items.

- How to compute the Kruskal–Wallis H-test.
- How to perform contrasts to compare samples.
- How to perform the Kruskal–Wallis H-test and associated sample contrasts using SPSS.

6.2 INTRODUCTION

A professor asked her students to complete end-of-course evaluations for her psychology 101 class. She taught four sections of the course and wants to compare the evaluation results from each section. Since the evaluations were based upon a five-point rating scale, she decides to use a nonparametric procedure. Moreover, she recognizes that the four sets of evaluation results are independent, or unrelated. In other words, no single score in any single class is dependent upon any other score in any other class. This professor could compare her sections using the Kruskal–Wallis H-test.

Nonparametric Statistics for Non-Statisticians, Gregory W. Corder and Dale I. Foreman
Copyright © 2009 John Wiley & Sons, Inc.

The Kruskal–Wallis H-test is a nonparametric statistical procedure for comparing more than two samples that are independent, or not related. The parametric equivalent to this test is the one-way analysis of variance (ANOVA).

When the Kruskal–Wallis H-test leads to significant results, then at least one of the samples is different from the other samples. However, the test does not identify where the difference(s) occur. Moreover, it does not identify how many differences occur. To identify the particular differences between sample pairs, a researcher might use sample contrasts, or post hoc tests, to analyze the specific sample pairs for significant difference(s). The Mann–Whitney U-test is a useful method for performing sample contrasts between individual sample sets.

In this chapter, we will describe how to perform and interpret a Kruskal–Wallis H-test followed with sample contrasts. We will also explain how to perform the procedures using SPSS. Finally, we offer varied examples of these nonparametric statistics from the literature.

6.3 COMPUTING THE KRUSKAL–WALLIS H-TEST STATISTIC

The Kruskal–Wallis H-test is used to compare more than two independent samples. When stating our hypotheses, we state them in terms of the population. Moreover, we examine the population medians, θ_i, when performing the Kruskal–Wallis H-test.

To compute the Kruskal–Wallis H-test statistic, we begin by combining all of the samples and rank ordering the values together. Use Formula 6.1 to determine an H statistic.

$$H = \frac{12}{N(N+1)} \sum_{i=1}^{k} \frac{R_i^2}{n_i} - 3(N+1) \tag{6.1}$$

where N is the number of values from all combined samples, R_i is the sum of the ranks from a particular sample, and n_i is the number of values from the corresponding rank sum.

The degrees of freedom, df, for the Kruskal–Wallis H-test are determined by using Formula 6.2.

$$\text{df} = k-1 \tag{6.2}$$

where df is the degrees of freedom and k is the number of groups.

Once the test statistic, H, is computed, it can be compared to a table of critical values (see Table B.6) to examine the groups for significant differences. However, if the number of groups, k, or the number of values in each sample, n_i, exceeds those available from the table, then a large sample approximation may be performed. Use a table with the chi-square distribution (see Table B.2) to obtain a critical value when performing a large sample approximation.

If ranking of values results in any ties, a ties correction is required. In that case, find a new H statistic by dividing the original H statistic by the ties correction. Use

Formula 6.3 to determine the ties correction value.

$$C_H = 1 - \frac{\sum(T^3 - T)}{N^3 - N} \quad (6.3)$$

where C_H is the ties correction, T is the number of values from a set of ties, and N is the number of values from all combined samples.

If the H statistic is not significant, then no differences exist between any of the samples. However, if the H statistic is significant, then a difference exists between at least two of the samples. Therefore, a researcher might use sample contrasts between individual sample pairs, or post hoc tests, to determine which of the sample pairs are significantly different.

When performing multiple sample contrasts, the Type I error rate tends to become inflated. Therefore, the initial level of risk, or α, must be adjusted. We recommend the Bonferroni procedure, shown in Formula 6.4, to adjust α.

$$\alpha_B = \frac{\alpha}{k} \quad (6.4)$$

where α_B is the adjusted level of risk, α is the original level of risk, and k is the number of comparisons.

6.3.1 Sample Kruskal–Wallis H-Test (Small Data Samples)

Researchers were interested in studying the social interaction of different adults. They sought to determine if social interaction can be tied to self-confidence. The researchers classified 17 participants into three groups based on the social interaction exhibited. The participant groups were labeled

High: constant interaction; talks with many different people; initiates discussion.
Medium: interacts with a variety of people; some periods of isolation; tends to focus on fewer people.
Low: remains mostly isolated from others; speaks if spoken to, but leaves interaction quickly.

After the participants had been classified into the three social interaction groups, they were directed to complete a self-assessment of self-confidence on a 25-point scale. Table 6.1 shows the scores obtained by each of the participants, with 25 points being an indication of high self-confidence.

The original survey scores obtained were converted to an ordinal scale prior to the data analysis. Table 6.1 shows the ordinal values placed in the social interaction groups.

We want to determine if there is a difference between any of the three groups in Table 6.1. Since the data belong to an ordinal scale and the sample sizes are small ($n < 20$), we require a nonparametric test. The Kruskal–Wallis H-test is the best statistic to analyze the data and test the hypothesis.

TABLE 6.1

Original Ordinal Self-Confidence Scores Placed Within Social Interaction Groups

High	Medium	Low
21	19	7
23	5	8
18	10	15
12	11	3
19	9	6
20		4

1. *State the null and research hypotheses.*

    The null hypothesis, shown below, states that there is no tendency for self-confidence to rank systematically higher or lower for any of the levels of social interaction. The research hypothesis states that there is a tendency for self-confidence to rank systematically higher or lower for at least one level of social interaction compared to at least one of the other levels. We generally use the concept of "systematic differences" in the hypotheses.

    The null hypothesis is

    H_0: $\theta_L = \theta_M = \theta_H$

    The research hypothesis is

    H_A: There is a tendency for self-confidence to rank systematically higher or lower for at least one level of social interaction when compared to the other levels.

2. *Set the level of risk (or the level of significance) associated with the null hypothesis.*

    The level of risk, also called an alpha (α), is frequently set at 0.05. We will use an alpha of 0.05 in our example. In other words, there is a 95% chance that any observed statistical difference will be real and not due to chance.

3. *Choose the appropriate test statistic.*

    The data are obtained from three independent, or unrelated, samples of adults who are being assigned to three different social interaction groups by observation. They are then being assessed using a self-confidence scale with a total of 25 points. The three samples are small with some violations of our assumptions of normality. Since we are comparing three independent samples, we will use the Kruskal–Wallis *H*-test.

4. *Compute the test statistic.*

    First, combine and rank the three samples together (see Table 6.2). Place the participant ranks in their social interaction groups to compute the sum of ranks, R_i, for each group (see Table 6.3). Next, compute the sum of ranks for each social interaction group. The ranks in each group are added to obtain a total R value for the group.

COMPUTING THE KRUSKAL–WALLIS H-TEST STATISTIC

TABLE 6.2

Original Ordinal Score	Participant Rank	Social Interaction Group
3	1	Low
4	2	Low
5	3	Medium
6	4	Low
7	5	Low
8	6	Low
9	7	Medium
10	8	Medium
11	9	Medium
12	10	High
15	11	Low
18	12	High
19	13.5	Medium
19	13.5	High
20	15	High
21	16	High
23	17	High

For the High group,

$$R_H = 10 + 12 + 13.5 + 15 + 16 + 17 = 83.5$$
$$n_H = 6$$

For the Medium group,

$$R_M = 3 + 7 + 8 + 9 + 13.5 = 40.5$$
$$n_M = 5$$

For the Low group,

$$R_L = 1 + 2 + 4 + 5 + 6 + 11 = 29$$
$$n_L = 6$$

TABLE 6.3

	Ordinal Data Ranks		
High	Medium	Low	
10	3	1	$N = 17$
12	7	2	
13.5	8	4	
15	9	5	
16	13.5	6	
17		11	

These R values are used to compute the Kruskal–Wallis H-test statistic (see Formula 6.1). The number of participants in each group is identified by a lowercase n. The total group size in the study is identified by the uppercase N.

Now, using the data from Table 6.3, compute the H-test statistic using Formula 6.1.

$$H = \frac{12}{N(N+1)} \sum_{i=1}^{k} \frac{R_i^2}{n_i} - 3(N+1)$$

$$= \frac{12}{17(17+1)} \left(\frac{83.5^2}{6} + \frac{40.5^2}{5} + \frac{29^2}{6} \right) - 3(17+1)$$

$$= 0.0392 \, (1162.04 + 328.05 + 140.17) - 54$$

$$= 0.0392 \, (1630.26) - 54$$

$$= 63.93 - 54$$

$$= 9.93$$

Since there was a tie involved in the ranking, correct the value of H. First, compute the ties correction (see Formula 6.2). Then, divide the original H statistic by the ties correction, C_H.

$$C_H = 1 - \frac{\sum (T^3 - T)}{N^3 - N}$$

$$= 1 - \frac{(2^3 - 2)}{17^3 - 17}$$

$$= 1 - \frac{(8 - 2)}{(4913 - 17)}$$

$$= 1 - 0.0001$$

$$= 0.9988$$

Next, we divide to find the corrected H statistic.

$$\text{corrected } H = \text{original } H / C_H = 9.93/0.9988 = 9.94$$

For this set of data, notice that the corrected H does not differ greatly from the original H. With the correction, $H = 9.94$.

5. *Determine the value needed for rejection of the null hypothesis using the appropriate table of critical values for the particular statistic.*

    We will use the critical values table for the Kruskal–Wallis H-test (see Table B.6) since it includes the number of groups, k, and the number of samples, n, for our data. In this case, we look for the critical value for $k = 3$ and $n_1 = 6, n_2 = 6$, and $n_3 = 5$ with $\alpha = 0.05$. Table B.5 returns a critical value for the Kruskal–Wallis H-test of 5.76.

6. *Compare the obtained value to the critical value.*

    The critical value for rejecting the null hypothesis is 5.76 and the obtained value is $H = 9.94$. If the critical value is less than or equal to the obtained value, we must reject the null hypothesis. If instead, the critical value exceeds the obtained value, we do not reject the null hypothesis. Since the critical value is less than the obtained value, we must reject the null hypothesis.

At this point, it is worth mentioning that larger samples often result in more ties. While comparatively small, as observed in step 4, corrections for ties can make a difference in the decision regarding the null hypothesis. If the H were near the critical value of 5.99 for df $= 2$ (e.g., $H = 5.80$), and the ties correction calculated to be 0.965, the decision would be to reject the null hypothesis with the correction ($H = 6.01$), but to not reject the null hypothesis without the correction. Therefore, it is important to perform ties corrections.

7. *Interpret the results.*

We rejected the null hypothesis, suggesting that a real difference in self-confidence exists between one or more of the three social interaction types. In particular, the data show that those who were classified as fitting the definition of the "Low" group were mostly people who reported poor self-confidence and those who were in the "High" group were mostly people who reported good self-confidence. However, describing specific differences in this manner is speculative. Therefore, we need a technique for statistically identifying difference between groups, or contrasts.

7a. *Sample contrasts, or post hoc tests.*

The Kruskal–Wallis H-test identifies if a statistical difference exists; however, it does not identify how many differences exist and which samples are different. To identify which samples are different and which are not, we can use a procedure called contrasts, or post hoc tests. An appropriate test to use when comparing two samples at a time is the Mann–Whitney U-test described in Chapter 3.

It is important to note that performing several Mann–Whitney U-tests has a tendency to inflate the Type I error rate. In our example, we would compare three groups, $k = 3$. At an $\alpha = 0.05$, the Type I error rate would equal $1 - (1 - 0.05)^3 = 0.14$.

To compensate for this error inflation, we suggest using the Bonferroni procedure (see Formula 6.4). With this technique, we use a corrected α with the Mann–Whitney U-tests to determine significant differences between samples. For our example,

$$\alpha_B = \frac{\alpha}{k} = \frac{0.05}{3} = 0.0167$$

When we compare the three samples with the Mann–Whitney U-tests and α_B (see Chapter 3), we obtain the following results presented in Table 6.4.

Using $\alpha_B = 0.0167$, we notice that the High–Low group comparison is indeed significantly different. The Medium–Low group comparison is not significant. The High–Medium group comparison requires some judgment since it is difficult to tell if the difference is significant or not; the way the value is rounded could change the result.

TABLE 6.4

Group Comparison	Mann–Whitney U Statistic	Rank Sum Difference	Significance
High–Medium	2.5	48.5 – 17.5 = 31.0	0.017
Medium–Low	7.0	38.0 – 28.0 = 10.0	0.177
High–Low	1.0	56.0 – 22.0 = 34.0	0.004

Note: If you are not comparing all of the samples for the Kruskal–Wallis *H*-test, then k is only the number of comparisons you are making with the Mann–Whitney *U*-tests. Therefore, comparing fewer samples will increase the chances of finding a significant difference.

8. *Reporting the results.*

   The reporting of results for the Kruskal–Wallis *H*-test should include such information as sample size for all of the groups, the *H* statistic, degrees of freedom, and *p*-value's relation to α. For this example, three social interaction groups were compared: high ($n_H = 6$), medium ($n_M = 5$), and low ($n_L = 6$). The Kruskal–Wallis *H*-test was significant ($H_{(2)} = 9.94$, $p < 0.05$). To compare individual pairs of samples, contrasts may be used (see Chapter 3).

6.3.2 Performing the Kruskal–Wallis *H*-Test Using SPSS

We will analyze the data from the above example using SPSS.

1. *Define your variables.*

   First, click the "Variable View" tab at the bottom of your screen. Then, type the names of your variables in the "Name" column. Unlike the Friedman ANOVA described in Chapter 4, you cannot simply enter each sample into a separate column to execute the Kruskal–Wallis *H*-test. You must use a grouping variable. In Figure 6.1, the first variable is the grouping variable that we called "Group". The second variable that we called "Score" will have our actual values.

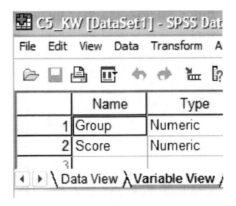

FIGURE 6.1

COMPUTING THE KRUSKAL–WALLIS H-TEST STATISTIC

When establishing a grouping variable, it is often easiest to assign each group a whole number value. In our example, our groups are "High", "Medium", and "Low". Therefore, we must set our grouping variables for the variable "Group". First, we selected the "Values" column and clicked the gray square as shown in Figure 6.2. Then, we set a value of 1 to equal "High" and a value of 2 to equal "Medium". As soon as we click the "Add" button, we will have set "Low" equal to 3.

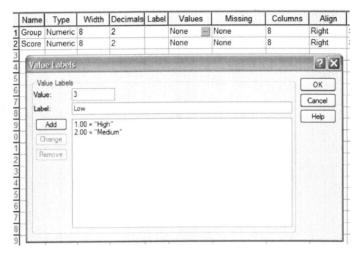

FIGURE 6.2

2. *Type in your values.*

   Click the "Data View" tab at the bottom of your screen as shown in Figure 6.3. Type in the values for all three samples in the "Score" column. As you do so, type in the corresponding grouping variable in the "Group" column. For example, all of the values for "High" are signified by a value of 1 in the grouping variable column that we called "Group".

3. *Analyze your data.*

   As shown in Figure 6.4, use the pull-down menus to choose "Analyze", "Nonparametric Tests", and "K Independent Samples…".

   Use the top arrow button to place your variable with your data values, or dependent variable (DV), in the box labeled "Test Variable List:". Then, use the lower arrow button to place your grouping variable, or independent variable (IV), in the box labeled "Grouping Variable". As shown in Figure 6.5, we have placed the "Score" variable in the "Test Variable List" and the "Group" variable in the "Grouping Variable" box.

   Click on the "Define Range…" button to assign a reference value to your independent variable (i.e., "Grouping Variable").

	Group	Score
1	1.00	21.00
2	1.00	23.00
3	1.00	18.00
4	1.00	12.00
5	1.00	19.00
6	1.00	20.00
7	2.00	19.00
8	2.00	5.00
9	2.00	10.00
10	2.00	11.00
11	2.00	9.00
12	3.00	7.00
13	3.00	8.00
14	3.00	15.00
15	3.00	3.00
16	3.00	6.00
17	3.00	4.00

FIGURE 6.3

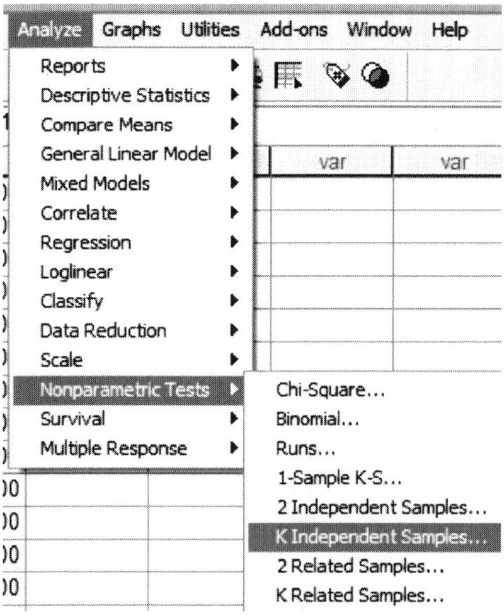

FIGURE 6.4

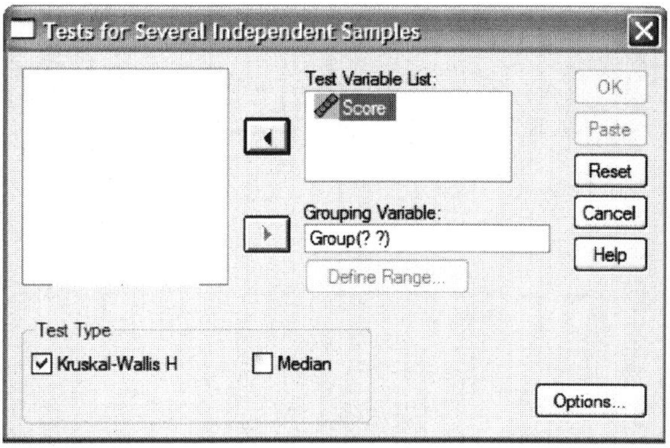

COMPUTING THE KRUSKAL–WALLIS *H*-TEST STATISTIC

FIGURE 6.5

As shown in Figure 6.6, type 1 into the box next to "Minimum" and 3 in the box next to "Maximum". Then, click "Continue". This step references the value labels you defined when you established your grouping variable.

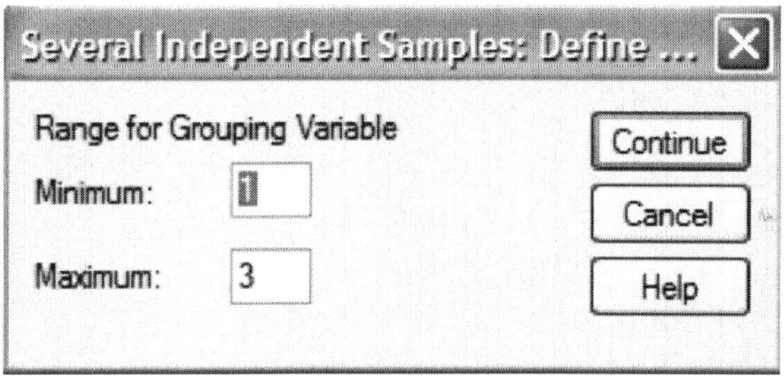

FIGURE 6.6

Now that the groups have been assigned (see Figure 6.7), click "OK" to perform the analysis.

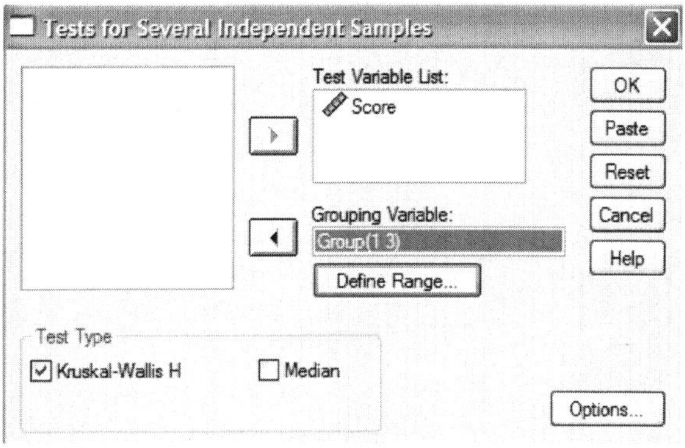

FIGURE 6.7

4. *Interpret the results from the SPSS Output window.*

Ranks

	Group	N	Mean Rank
Score	High	6	13.92
	Medium	5	8.10
	Low	6	4.83
	Total	17	

Test Statistics[a,b]

	Score
Chi-Square	9.944
df	2
Asymp. Sig.	.007

a. Kruskal Wallis Test
b. Grouping Variable: Group

The first SPSS output table provides the mean ranks of groups and group sizes. The SPSS second output table provides the Kruskal–Wallis H-test statistic ($H = 9.944$). Since this test uses a chi-square distribution, SPSS calls the H statistic "Chi-Square". This table also returns the degrees of freedom (df $= 2$) and the significance ($p = 0.007$).

Based on the results from SPSS, three social interaction groups were compared: high ($n_H = 6$), medium ($n_M = 5$), and low ($n_L = 6$). The Kruskal–Wallis H-test was significant ($H_{(2)} = 9.94$, $p < 0.05$). To compare individual pairs of samples, contrasts must be used.

Note: To perform Mann–Whitney U-tests for sample contrasts, simply use the grouping values you established when you defined your variables in step 1. Remember to use your corrected level of risk, α_B, when examining your significance.

6.3.3 Sample Kruskal–Wallis H-Test (Large Data Samples)

Researchers were interested in continuing their study of social interaction. In a new study, they examined the self-confidence of teenagers with respect to social interaction. Three levels of social interaction were based upon the following characteristics.

High: constant interaction; talks with many different people; initiates discussion.

Medium: interacts with a variety of people; some periods of isolation; tends to focus on fewer people.

Low: remains mostly isolated from others; speaks if spoken to, but leaves interaction quickly.

COMPUTING THE KRUSKAL–WALLIS H-TEST STATISTIC

The researchers assigned each participant into one of the three social interaction groups. Researchers administered a self-assessment of self-confidence. The assessment instrument measured self-confidence on a 50-point ordinal scale. Table 6.5 shows the scores obtained by each of the participants, with 50 points indicating high self-confidence.

We want to determine if there is a difference between any of the three groups in Table 6.5. The Kruskal–Wallis H-test will be used to analyze the data.

TABLE 6.5

Original Self-Confidence Scores Placed Within Social Interaction Groups

High	Medium	Low
18	35	37
27	47	24
24	11	7
30	31	19
48	12	20
16	39	14
43	11	38
46	14	16
49	40	12
34	48	31
28	32	15
20	9	20
37	44	25
21	30	10
20	33	36
16	26	45
23	22	48
12	3	42
50	41	42
25	17	21
	8	
	10	
	41	

1. *State the null and research hypotheses.*

   The null hypothesis, shown below, states that there is no tendency for teen self-confidence to rank systematically higher or lower for any of the levels of social interaction. The research hypothesis states that there is a tendency for teen self-confidence to rank systematically higher or lower for at least one level of social interaction compared to at least one of the other levels. We generally use the concept of "systematic differences" in the hypotheses.

The null hypothesis is

H$_0$: $\theta_L = \theta_M = \theta_H$

The research hypothesis is

H$_A$: There is a tendency for teen self-confidence to rank systematically higher or lower for at least one level of social interaction when compared to the other levels.

2. *Set the level of risk (or the level of significance) associated with the null hypothesis.*

   The level of risk, also called an alpha (α), is frequently set at 0.05. We will use an alpha of 0.05 in our example. In other words, there is a 95% chance that any observed statistical difference will be real and not due to chance.

3. *Choose the appropriate test statistic.*

   The data are obtained from three independent, or unrelated, samples of teenagers. They were assessed using an instrument with a 50-point ordinal scale. Since we are comparing three independent samples of values based on an ordinal scale instrument, we will use the Kruskal–Wallis *H*-test.

4. *Compute the test statistic.*

   First, combine and rank the three samples together (see Table 6.6).

TABLE 6.6

Original Ordinal	Participant Score Rank	Social Interaction Group
3	1	Medium
7	2	Low
8	3	Medium
9	4	Medium
10	5.5	Medium
10	5.5	Low
11	7.5	Medium
11	7.5	Medium
12	10	High
12	10	Medium
12	10	Low
14	12.5	Medium
14	12.5	Low
15	14	Low
16	16	High
16	16	High
16	16	Low
17	18	Medium
18	19	High
19	20	Low
20	22.5	High
20	22.5	High
20	22.5	Low

COMPUTING THE KRUSKAL–WALLIS H-TEST STATISTIC

TABLE 6.6 (*Continued*)

Original Ordinal	Participant Score Rank	Social Interaction Group
20	22.5	Low
21	25.5	High
21	25.5	Low
22	27	Medium
23	28	High
24	29.5	High
24	29.5	Low
25	31.5	High
25	31.5	Low
26	33	Medium
27	34	High
28	35	High
30	36.5	High
30	36.5	Medium
31	38.5	Medium
31	38.5	Low
32	40	Medium
33	41	Medium
34	42	High
35	43	Medium
36	44	Low
37	45.5	High
37	45.5	Low
38	47	Low
39	48	Medium
40	49	Medium
41	50.5	Medium
41	50.5	Medium
42	52.5	Low
42	52.5	Low
43	54	High
44	55	Medium
45	56	Low
46	57	High
47	58	Medium
48	60	High
48	60	Medium
48	60	Low
49	62	High
50	63	High

TABLE 6.7

	Ordinal Data Ranks	
High	Medium	Low
10	1	2
16	3	5.5
16	4	10
19	5.5	12.5
22.5	7.5	14
22.5	7.5	16
25.5	10	20
28	12.5	22.5
29.5	18	22.5
31.5	27	25.5
34	33	29.5
35	36.5	31.5
36.5	38.5	38.5
42	40	44
45.5	41	45.5
54	43	47
57	48	52.5
60	49	52.5
62	50.5	56
63	50.5	60
	55	
	58	
	60	

Place the participant ranks in their social interaction groups to compute the sum of ranks, R_i, for each group (see Table 6.7).

Next, compute the sum of ranks for each social interaction group. The ranks in each group are added to obtain a total R value for the group.

For the High group, $R_H = 709.5$ and $n_H = 20$.
For the Medium group, $R_M = 699$ and $n_M = 23$.
For the Low group, $R_L = 607.5$ and $n_L = 20$.

These R values are used to compute the Kruskal–Wallis H-test statistic (see Formula 6.1). The number of participants in each group is identified by a lowercase n. The total group size in the study is identified by the uppercase N. In this study, $N = 63$.

Now, using the data from Table 6.7, compute the H-test statistic using Formula 6.1.

$$H = \frac{12}{N(N+1)} \sum_{i=1}^{k} \frac{R_i^2}{n_i} - 3(N+1)$$

$$= \frac{12}{63(63+1)} \left(\frac{709.5^2}{20} + \frac{699^2}{23} + \frac{607.5^2}{20} \right) - 3(63+1)$$

$$= 0.003 \, (25169.51 + 21243.52 + 18452.81) - 192$$

$$= 0.003 \, (64865.85) - 192$$

$$= 1.053$$

Since there were ties involved in the ranking, correct the value of H. First, compute the ties correction (see Formula 6.2). There were 11 sets of ties with two values, 3 sets of ties with three values, and 1 set of ties with four values. Then, divide the original H statistic by the ties correction, C_H.

$$C_H = 1 - \frac{\sum (T^3 - T)}{N^3 - N}$$

$$= 1 - \frac{11(2^3 - 2) + 3(3^3 - 3) + (4^3 - 4)}{63^3 - 63}$$

$$= 1 - \frac{189}{249984}$$

$$= 1 - 0.0008$$

$$= 0.9992$$

Next, we divide to find the corrected H statistic.

corrected H = original H/C_H = $1.053/0.9992$

For this set of data, notice that the corrected H does not differ greatly from the original H. With the correction, $H = 1.054$.

5. *Determine the value needed for rejection of the null hypothesis using the appropriate table of critical values for the particular statistic.*

   Since the data have at least one large sample, we will use the chi-square distribution (see Table B.2) to find the critical value for the Kruskal–Wallis H-test. In this case, we look for the critical value for df = 2 and $\alpha = 0.05$. Using the table, the critical value for rejecting the null hypothesis is 5.99.

6. *Compare the obtained value to the critical value.*

   The critical value for rejecting the null hypothesis is 5.99 and the obtained value is $H = 1.054$. If the critical value is less than or equal to the obtained value, we must reject the null hypothesis. If instead, the critical value exceeds the obtained value, we do not reject the null hypothesis. Since the critical value exceeds the obtained value, we do not reject the null hypothesis.

7. *Interpret the results.*

We did not reject the null hypothesis, suggesting that no real difference exists between any of the three groups. In particular, the data suggest that there is no difference in self-confidence between one or more of the three social interaction types.

8. *Reporting the results.*

The reporting of results for the Kruskal–Wallis H-test should include such information as sample size for each of the groups, the H statistic, degrees of freedom, and p-value's relation to α. For this example, three social interaction groups were compared. The three social interaction groups were high ($n_H = 20$), medium ($n_M = 23$), and low ($n_L = 20$). The Kruskal–Wallis H-test was not significant ($H_{(2)} = 1.054$, $p > 0.05$).

6.4 EXAMPLES FROM THE LITERATURE

Below are varied examples of the nonparametric procedures described in this chapter. We have summarized each study's research problem and researchers' rationale(s) for choosing a nonparametric approach. We encourage you to obtain these studies if you are interested in their results.

- Gömleksız, M. N., & Bulut, İ. (2007). An evaluation of the effectiveness of the new primary school mathematics curriculum in practice. *Educational Sciences: Theory & Practice, 7*(1), 81–94.

    Gömleksız and Bulut examined primary school teachers' views on the implementation and effectiveness of a new primary school mathematics curriculum. When they examined the data, some of the samples were found to be non-normal. For those samples, they used a Kruskal–Wallis H-test, followed by Mann–Whitney U-tests to compare unrelated samples.

- Finson, K. D., Pedersen, J., & Thomas, J. (2006). Comparing science teaching styles to students' perceptions of scientists. *School Science and Mathematics, 106*(1), 8–15.

    In Finson, Pedersen, and Thomas' study, the students of nine middle school teachers were asked to draw a scientist. Based on the drawings, students' perceptions of scientists were compared with their teachers' teaching styles using the Kruskal–Wallis H-test. Then, the samples were individually compared using the Mann–Whitney U-test. The researchers used nonparametric statistical analyses because only relatively small sample sizes of subjects were available.

- Belanger, N. D., & Desrochers, S. (2001). Can 6-month-old infants process causality in different types of causal events? *British Journal of Developmental Psychology, 19*(1), 11–21.

    Belanger and Desrochers investigated the nature of infants' ability to perceive event causality. The researchers noted that they chose nonparametric

statistical tests because the data samples lacked a normal distribution based on results from a Shapiro–Wilk test. In addition, they stated that the sample sizes were small. A Kruskal–Wallis H-test revealed no significant differences between samples. Therefore, they did not perform any sample contrasts.
- Plata, M., & Trusty, J. (2005). Effect of socioeconomic status on general and at-risk high school boys' willingness to accept same-sex peers with LD. *Adolescence, 40*(157), 47–66.

  Plata and Trusty investigated high school boys' willingness to allow same-sex peers with learning disabilities (LD) to participate in school activities and out-of-school activities. The authors compared the willingness of 38 educationally successful and 33 educationally at-risk boys. The boys were from varying socioeconomic backgrounds. Due to the ordinal nature of data and small sample sizes among some samples, nonparametric statistics were used for the analysis. The Kruskal–Wallis H-test was chosen for multiple comparisons. When sample pairs were compared, the researchers performed a post hoc analysis of the differences between mean rank pairs using a multiple comparison technique.

6.5 SUMMARY

More than two samples that are not related may be compared using a nonparametric procedure called the Kruskal–Wallis H-test. The parametric equivalent to this test is known as the one-way analysis of variance. When the Kruskal–Wallis H-test produces significant results, it does not identify which or how many sample pairs are significantly different. The Mann–Whitney U-test, with a Bonferroni procedure to avoid Type I error rate inflation, is a useful method for comparing individual sample pairs.

In this chapter, we described how to perform and interpret a Kruskal–Wallis H-test followed with sample contrasts. We also explained how to perform the procedures using SPSS. Finally, we offered varied examples of these nonparametric statistics from the literature. The next chapter will involve comparing two variables.

6.6 PRACTICE QUESTIONS

1. A researcher conducted a study with $n = 15$ participants to investigate strength gains from exercise. The participants were divided into three groups and given one of three treatments. Participants' strength gains were measured and ranked. The rankings are presented in Table 6.8.
Use a Kruskal–Wallis H-test with $\alpha = 0.05$ to determine if one or more of the groups are significantly different. If a significant difference exists, use two-tailed Mann–Whitney U-tests to identify which groups are significantly different. Use the Bonferroni procedure to limit the Type I error rate. Report your findings.

TABLE 6.8

	Treatments	
I	II	III
7	13	12
2	1	5
4	7	16
11	8	9
15	3	14

2. A researcher investigated how physical attraction influences the perception among others of a person's effectiveness with difficult tasks. The photographs of 24 people were shown to a focus group. The group was asked to classify the photos into three groups: very attractive, average, and very unattractive. Then, the group ranked the photographs according to their impression of how capable they were of solving difficult problems. Table 6.9 shows the classification and rankings of the people in the photos (1: most effective; 24: least effective).

TABLE 6.9

Very Attractive	Average	Very Unattractive
1	3	11
2	4	15
5	8	16
6	9	18
7	13	20
10	14	21
12	19	23
17	22	24

Use a Kruskal–Wallis H-test with $\alpha = 0.05$ to determine if one or more of the groups are significantly different. If a significant difference exists, use two-tailed Mann–Whitney U-tests to identify which groups are significantly different. Use the Bonferroni procedure to limit the Type I error rate. Report your findings.

6.7 SOLUTIONS TO PRACTICE QUESTIONS

1. The results from the Kruskal–Wallis H-test are displayed in the SPSS Outputs below.

SOLUTIONS TO PRACTICE QUESTIONS

Ranks

	Treatment	N	Mean Rank
RankGain	Treatment 1	5	7.40
	Treatment 2	5	6.00
	Treatment 3	5	10.60
	Total	15	

Test Statistics[a,b]

	RankGain
Chi-Square	2.800
df	2
Asymp. Sig.	.247

a. Kruskal Wallis Test
b. Grouping Variable: Treatment

According to the data, the results from the Kruskal–Wallis H-test indicated that the three groups are not significantly different ($H_{(2)} = 2.800$, $p > 0.05$). Therefore, no follow-up contrasts are needed.

2. The results from the Kruskal–Wallis H-test are displayed in the SPSS Outputs below.

Ranks

	Classification	N	Mean Rank
Ranking	Very Attractive	8	7.50
	Average	8	11.50
	Very Unattractive	8	18.50
	Total	24	

Test Statistics[a,b]

	Ranking
Chi-Square	9.920
df	2
Asymp. Sig.	.007

a. Kruskal Wallis Test
b. Grouping Variable: Classification

According to the data, the results from the Kruskal–Wallis H-test indicated that one or more of the three groups are significantly different ($H_{(2)} = 9.920$, $p < 0.05$). Therefore, we must examine each set of samples with follow-up contrasts to find the differences between groups.

Based on the significance from the Kruskal–Wallis H-test, we compare the samples with Mann–Whitney U-tests. Since there are $k = 3$ groups, use $\alpha_B = 0.0167$ to avoid Type I error rate inflation. The results from the Mann–Whitney U-tests are displayed in the SPSS Outputs below

a. *Very attractive–attractive comparison*

Ranks

	Classification	N	Mean Rank	Sum of Ranks
Ranking	Very Attractive	8	7.00	56.00
	Average	8	10.00	80.00
	Total	16		

Test Statistics[b]

	Ranking
Mann-Whitney U	20.000
Wilcoxon W	56.000
Z	-1.260
Asymp. Sig. (2-tailed)	.208
Exact Sig. [2*(1-tailed Sig.)]	.234[a]

a. Not corrected for ties.
b. Grouping Variable: Classification

The results from the Mann–Whitney U-test ($U = 20.0$, $n_1 = 8$, $n_2 = 8$, $p > 0.0167$) indicated that the two samples were not significantly different.

b. *Attractive–very unattractive comparison*

Ranks

	Classification	N	Mean Rank	Sum of Ranks
Ranking	Average	8	6.00	48.00
	Very Unattractive	8	11.00	88.00
	Total	16		

Test Statistics[b]

	Ranking
Mann-Whitney U	12.000
Wilcoxon W	48.000
Z	-2.100
Asymp. Sig. (2-tailed)	.036
Exact Sig. [2*(1-tailed Sig.)]	.038[a]

a. Not corrected for ties.
b. Grouping Variable: Classification

The results from the Mann–Whitney U-test ($U = 12.0$, $n_1 = 8$, $n_2 = 8$, $p > 0.0167$) indicated that the two samples were not significantly different.

c. *Very attractive–very unattractive comparison*

Ranks

	Classification	N	Mean Rank	Sum of Ranks
Ranking	Very Attractive	8	5.00	40.00
	Very Unattractive	8	12.00	96.00
	Total	16		

Test Statistics[b]

	Ranking
Mann-Whitney U	4.000
Wilcoxon W	40.000
Z	-2.941
Asymp. Sig. (2-tailed)	.003
Exact Sig. [2*(1-tailed Sig.)]	.002[a]

a. Not corrected for ties.
b. Grouping Variable: Classification

The results from the Mann–Whitney U-test ($U = 4.0$, $n_1 = 8$, $n_2 = 8$, $p < 0.0167$) indicated that the two samples were significantly different.

7

COMPARING VARIABLES OF ORDINAL OR DICHOTOMOUS SCALES: SPEARMAN RANK-ORDER, POINT-BISERIAL, AND BISERIAL CORRELATIONS

7.1 OBJECTIVES

In this chapter, you will learn the following items.

- How to compute the Spearman rank-order correlation coefficient.
- How to perform the Spearman rank-order correlation using SPSS.
- How to compute the point-biserial correlation coefficient.
- How to perform the point-biserial correlation using SPSS.
- How to compute the biserial correlation coefficient.

7.2 INTRODUCTION

The statistical procedures in this chapter are quite different from those in the last several chapters. Unlike this chapter, we had compared samples of data. This chapter, however, examines the relationship between two variables. In other words, this chapter will address how one variable changes with respect to another.

The relationship between two variables can be compared with a correlation analysis. If any of the variables is ordinal or dichotomous, we can use a nonparametric correlation. The Spearman rank-order correlation, also called the Spearman's rho, is used to compare the relationship between ordinal, or rank-ordered, variables. The

---

Nonparametric Statistics for Non-Statisticians, Gregory W. Corder and Dale I. Foreman
Copyright © 2009 John Wiley & Sons, Inc.

point-biserial and biserial correlations are used to compare the relationship between two variables if one of the variables is dichotomous. The parametric equivalent to these correlations is the Pearson product-moment correlation.

In this chapter, we will describe how to perform and interpret Spearman rank-order, point-biserial, and biserial correlations. We will also explain how to perform the procedures using SPSS. Finally, we offer varied examples of these nonparametric statistics from the literature.

7.3 THE CORRELATION COEFFICIENT

When comparing two variables, we use an obtained value called a correlation coefficient. A population's correlation coefficient is represented by the Greek letter rho, ρ. A sample's correlation coefficient is represented by the letter r.

We will describe two types of relationships between variables. A direct relationship is a positive correlation with an obtained value ranging from 0 to 1.0. As one variable increases, the other variable also increases. An indirect, or inverse, relationship is a negative correlation with an obtained value ranging from 0 to -1.0. In this case, one variable increases as the other variable decreases.

In general, a significant correlation coefficient also communicates the relative strength of a relationship between the two variables. A value close to 1.0 or -1.0 indicates a nearly perfect relationship, while a value close to 0 indicates an especially weak or trivial relationship. Cohen (1988, 1992) presented a more detailed description of a correlation coefficient's relative strength. Table 7.1 displays his findings.

TABLE 7.1

Correlation Coefficient for a Direct Relationship	Correlation Coefficient for an Indirect Relationship	Relationship Strength of the Variables
0.0	0.0	None/trivial
0.1	-0.1	Weak/small
0.3	-0.3	Moderate/medium
0.5	-0.5	Strong/large
1.0	-1.0	Perfect

There are three important caveats to consider when assigning relative strength to correlation coefficients, however. First, Cohen's work was largely based on behavioral science research. Therefore, these values may be inappropriate in fields such as engineering or chemistry. Second, the correlation strength assignments vary for different types of statistical tests. Third, r values are not based on a linear scale. For example, $r = 0.6$ is not twice as strong as $r = 0.3$.

7.4 COMPUTING THE SPEARMAN RANK-ORDER CORRELATION COEFFICIENT

The Spearman rank-order correlation is a statistical procedure that is designed to measure the relationship between two variables on an ordinal scale of measurement if the sample size is $n \geq 4$. Use Formula 7.1 to determine a Spearman rank-order correlation coefficient, r_s, if none of the ranked values are ties. Sometimes, the symbol r_s is represented by the Greek symbol rho, or ρ.

$$r_s = 1 - \frac{6 \sum D_i^2}{n(n^2-1)} \quad (7.1)$$

where n is the number of rank pairs and D_i is the difference between a ranked pair. If ties are present in the values, use Formulas 7.2–7.4 to determine r_s.

$$r_s = \frac{(n^3-n) - 6 \sum D_i^2 - (T_x + T_y)/2}{\sqrt{(n^3-n)^2 - (T_x+T_y)(n^3-n) + T_x T_y}} \quad (7.2)$$

where

$$T_x = \sum_{i=1}^{g} (t_i^3 - t_i) \quad (7.3)$$

and

$$T_y = \sum_{i=1}^{g} (t_i^3 - t_i) \quad (7.4)$$

where g is the number of tied groups in that variable and t_i is the number of tied values in a tie group. If there are no ties in a variable, then $T = 0$.

Use Formula 7.5 to determine the degrees of freedom for the correlation.

$$\text{df} = n - 2 \quad (7.5)$$

where n is the number of paired values.

After r_s is determined, it must be examined for significance. Small samples allow one to reference a table of critical values, such as Table B.7. However, if the sample size, n, exceeds those available from the table, then a large sample approximation may be performed. For large samples, compute a z-score and use a table with the normal distribution (see Table B.1) to obtain a critical region of z-scores. Formula 7.6 may be used to find the z-score of a correlation coefficient for large samples.

$$z^* = r\left[\sqrt{n-1}\right] \quad (7.6)$$

where n is the number of paired values and r is the correlation coefficient.

COMPUTING THE SPEARMAN RANK-ORDER CORRELATION COEFFICIENT

Note: The method for determining a z-score given a correlation coefficient and examining it for significance is the same for each type of correlation. We will illustrate a large sample approximation with a sample problem when we address the point-biserial correlation.

Although we will use Formula 7.6 to determine the significance of the correlation coefficient, some statisticians recommend using the formula based on the Student's t distribution, as shown in Formula 7.7.

$$t = r_s \sqrt{\frac{n-2}{1-r_s^2}} \qquad (7.7)$$

According to Siegel and Castellan (1988), the advantage of using the Student's t distribution over the z-score is small with larger sample sizes, n.

7.4.1 Sample Spearman Rank-Order Correlation (Small Data Samples Without Ties)

Eight men were involved in a study to examine the resting heart rate regarding frequency of visits to the gym. The assumption is that the person who visits the gym more frequently for a workout will have a slower heart rate. Table 7.2 shows the number of visits each participant made to the gym during the month the study was conducted. It also provides the mean heart rate measured at the end of the week during the final 3 weeks of the month.

TABLE 7.2

Participant	Number of Visits	Mean Heart Rate
1	5	100
2	12	89
3	7	78
4	14	66
5	2	77
6	8	103
7	15	67
8	17	63

The values in this study do not possess characteristics of a strong interval scale. For instance, the number of visits to the gym does not necessarily communicate duration and intensity of physical activity. In addition, heart rate has several factors that can result in differences from one person to another. Ordinal measures offer a clearer relationship to compare these values from one individual to the next. Therefore, we will convert these values to ranks and use a Spearman rank-order correlation.

1. *State the null and research hypotheses.*

    The null hypothesis, shown below, states that there is no correlation between number of visits to the gym in a month and mean resting heart rate. The research hypothesis states that there is a correlation between the number of visits to the gym and the mean resting heart rate.

    The null hypothesis is

    $H_0: \rho_s = 0$

    The research hypothesis is

    $H_A: \rho_s \neq 0$

2. *Set the level of risk (or the level of significance) associated with the null hypothesis.*

    The level of risk, also called an alpha (α), is frequently set at 0.05. We will use an alpha of 0.05 in our example. In other words, there is a 95% chance that any observed statistical difference will be real and not due to chance.

3. *Choose the appropriate test statistic.*

    As stated earlier, we decided to analyze the variables using an ordinal, or rank, procedure. Therefore, we will convert the values in each variable to ordinal data. In addition, we will be comparing the two variables, number of visits to the gym in a month and mean resting heart rate. Since we are comparing two variables in which one or both are measured on an ordinal scale, we will use the Spearman rank-order correlation.

4. *Compute the test statistic.*

    First, rank the scores for each variable separately as shown in Table 7.3. Rank them from the lowest score to the highest score to form an ordinal distribution for each variable.

TABLE 7.3

Participant	Original Scores		Ranked Scores	
	Number of Visits	Mean Heart Rate	Number of Visits	Mean Heart Rate
1	5	100	2	7
2	12	89	5	6
3	7	78	3	5
4	14	66	6	2
5	2	77	1	4
6	8	103	4	8
7	15	67	7	3
8	17	63	8	1

To calculate the Spearman rank-order correlation coefficient, we need to calculate the differences between rank pairs and their subsequent squares where $D =$ rank(mean heart rate) $-$ rank(number of visits). It is helpful to organize the data to manage the summation in the formula (see Table 7.4).

TABLE 7.4

Ranked Scores		Rank Differences	
Number of Visits	Mean Heart Rate	D	D^2
2	7	5	25
5	6	1	1
3	5	2	4
6	2	−4	16
1	4	3	9
4	8	4	16
7	3	−4	16
8	1	−7	49
			$\sum D_i^2 = 136$

Next, compute the Spearman rank-order correlation coefficient.

$$r_s = 1 - \frac{6 \sum D_i^2}{n(n^2-1)}$$

$$= 1 - \frac{6(136)}{8(8^2-1)} = 1 - \frac{816}{8(64-1)}$$

$$= 1 - \frac{816}{8(63)} = 1 - \frac{816}{504}$$

$$= 1 - 1.619$$

$$= -0.619$$

5. *Determine the value needed for rejection of the null hypothesis using the appropriate table of critical values for the particular statistic.*

   Table B.7 lists critical values for the Spearman rank-order correlation coefficient. In this study, the critical value is found for $n=8$ and $df=6$. Since we are conducting a two-tailed test and $\alpha=0.05$, the critical value is 0.738. If the obtained value exceeds or is equal to the critical value, 0.738, we will reject the null hypothesis. If the critical value exceeds the absolute value of the obtained value, we will not reject the null hypothesis.

6. *Compare the obtained value to the critical value.*

   The critical value for rejecting the null hypothesis is 0.738 and the obtained value is $|r_s|=0.619$. If the critical value is less than or equal to the obtained value, we must reject the null hypothesis. If instead, the critical value is greater than the obtained value, we must not reject the null hypothesis. Since the critical value exceeds the absolute value of the obtained value, we do not reject the null hypothesis.

7. *Interpret the results.*

   We did not reject the null hypothesis, suggesting that there is no significant correlation between the number of visits the males made to the gym in a month and their mean resting heart rates.

128 COMPARING VARIABLES OF ORDINAL OR DICHOTOMOUS SCALES

8. *Reporting the results.*

   The reporting of results for the Spearman rank-order correlation should include such information as the number of participants (n), two variables that are being correlated, correlation coefficient (r_s), degrees of freedom (df), and p-value's relation to α.

   For this example, eight men ($n=8$) were observed for 1 month. Their number of visits to the gym were documented (variable 1) and their mean resting heart rate was recorded during the last 3 weeks of the month (variable 2). These data were put in ordinal form for purposes of the analysis. The Spearman rank-order correlation coefficient was not significant ($r_{s(6)} = -0.619$, $p > 0.05$). Based on these data, we can state that there is no clear relationship between adult male resting heart rate and the frequency of visits to the gym.

7.4.2 Sample Spearman Rank-Order Correlation (Small Data Samples with Ties)

The researcher repeated the experiment in the previous example using females. Table 7.5 shows the number of visits each participant made to the gym during the month of the study and their subsequent mean heart rates.

TABLE 7.5

Participant	Number of Visits	Mean Heart Rate
1	5	96
2	12	63
3	7	78
4	14	66
5	3	79
6	8	95
7	15	67
8	12	64
9	2	99
10	16	62
11	12	65
12	7	76
13	17	61

As with the previous example, the values in this study do not possess characteristics of a strong interval scale, so we will use ordinal measures. We will convert these values to ranks and use a Spearman rank-order correlation.

Steps 1–3 are the same as the previous example. Therefore, we will begin with step 4.

4. *Compute the test statistic.*

   First, rank the scores for each variable as shown in Table 7.6. Rank the scores from the lowest score to the highest score to form an ordinal distribution for each variable.

COMPUTING THE SPEARMAN RANK-ORDER CORRELATION COEFFICIENT

TABLE 7.6

Participant	Original Scores		Rank Scores	
	Number of Visits	Mean Heart Rate	Number of Visits	Mean Heart Rate
1	5	96	3	12
2	12	63	8	3
3	7	78	4.5	9
4	14	66	10	6
5	3	79	2	10
6	8	95	6	11
7	15	67	11	7
8	12	64	8	4
9	2	99	1	13
10	16	62	12	2
11	12	65	8	5
12	7	76	4.5	8
13	17	61	13	1

To calculate the Spearman rank-order correlation coefficient, we need to calculate the differences between rank pairs and their subsequent squares where $D = \text{rank(mean heart rate)} - \text{rank(number of visits)}$. It is helpful to organize the data to manage the summation in the formula (see Table 7.7).

Next, compute the Spearman rank-order correlation coefficient. Since there are ties present in the ranks, we will use formulas that account for the ties. First,

TABLE 7.7

Participant	Rank Scores		Rank Differences	
	Number of Visits	Mean Heart Rate	D	D^2
1	3	12	9	81
2	8	3	−5	25
3	4.5	9	4.5	20.25
4	10	6	−4	16
5	2	10	8	64
6	6	11	5	25
7	11	7	−4	16
8	8	4	−4	16
9	1	13	12	144
10	12	2	−10	100
11	8	5	−3	9
12	4.5	8	3.5	12.25
13	13	1	−12	144
				$\sum D_i^2 = 672.5$

use Formulas 7.3 and 7.4. For the number of visits, there are two groups of ties. The first group has two tied values (rank = 4.5 and $t = 2$) and the second group has three tied values (rank = 8 and $t = 3$).

$$T_x = \sum_{i=1}^{g}(t_i^3 - t_i)$$
$$= (2^3 - 2) + (3^3 - 3) = (8-2) + (27-3)$$
$$= 6 + 24$$
$$= 30$$

For the mean resting heart rate, there are no ties. Therefore, $T_y = 0$. Now, calculate the Spearman rank-order correlation coefficient using Formula 7.2.

$$r_s = \frac{(n^3-n) - 6\sum D_i^2 - (T_x + T_y)/2}{\sqrt{(n^3-n)^2 - (T_x+T_y)(n^3-n) + T_x T_y}}$$

$$= \frac{(13^3-13) - 6(672.5) - (30+0)/2}{\sqrt{(13^3-13)^2 - (30+0)(13^3-13) + (30)(0)}}$$

$$= \frac{2184 - 4035 - 15}{\sqrt{(2184)^2 - (30)(2184) + 0}}$$

$$= \frac{-1866}{\sqrt{4704336}} = \frac{-1866}{2169}$$

$$= -0.860$$

5. *Determine the value needed for rejection of the null hypothesis using the appropriate table of critical values for the particular statistic.*

    Table B.7 lists critical values for the Spearman rank-order correlation coefficient. To be significant, the absolute value of the obtained value, $|r_s|$, must be greater than or equal to the critical value on the table. In this study, the critical value is found for $n = 13$ and df = 11. Since we are conducting a two-tailed test and $\alpha = 0.05$, the critical value is 0.560.

6. *Compare the obtained value to the critical value.*

    The critical value for rejecting the null hypothesis is 0.560 and the obtained value is $|r_s| = 0.860$. If the critical value is less than or equal to the obtained value, we must reject the null hypothesis. If instead, the critical value is greater than the obtained value, we must not reject the null hypothesis. Since the critical value is less than the absolute value of the obtained value, we reject the null hypothesis.

7. *Interpret the results.*

    We rejected the null hypothesis, suggesting that there is a significant correlation between the number of visits the females made to the gym in a month and their mean resting heart rates.

8. *Reporting the results.*

The reporting of results for the Spearman rank-order correlation should include such information as the number of participants (n), two variables that are being correlated, correlation coefficient (r_s), degrees of freedom (df), and p-value's relation to α.

For this example, 13 women ($n = 13$) were observed for 1 month. Their number of visits to the gym were documented (variable 1) and their mean resting heart rate was recorded during the last 3 weeks of the month (variable 2). These data were put in ordinal form for purposes of the analysis. The Spearman rank-order correlation coefficient was significant ($r_{s(11)} = -0.860$, $p < 0.05$). Based on these data, we can state that there is a very strong inverse relationship between adult female resting heart rate and the frequency of visits to the gym.

7.4.3 Performing the Spearman Rank-Order Correlation Using SPSS

We will analyze the data from the previous example using SPSS.

1. *Define your variables.*

    First, click the "Variable View" tab at the bottom of your screen. Then, type the names of your variables in the "Name" column. As shown in Figure 7.1, the first variable is called "Number_of_Visits" and the second variable is called "Mean_Heart_Rate".

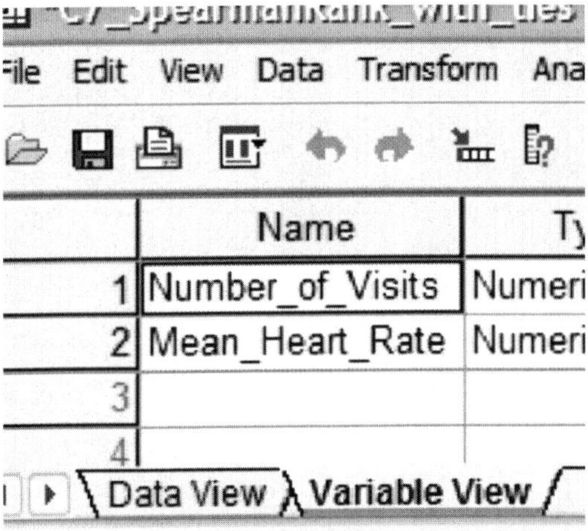

FIGURE 7.1

2. *Type in your values.*

    Click the "Data View" tab at the bottom of your screen as shown in Figure 7.2. Type the values in the respective columns.

COMPARING VARIABLES OF ORDINAL OR DICHOTOMOUS SCALES

	Number_of_Visits	Mean Heart Rate
1	5.00	96.00
2	12.00	63.00
3	7.00	78.00
4	14.00	66.00
5	3.00	79.00
6	8.00	95.00
7	15.00	67.00
8	12.00	64.00
9	2.00	99.00
10	16.00	62.00
11	12.00	65.00
12	7.00	76.00
13	17.00	61.00

\Data View /\ Variable View /

FIGURE 7.2

3. *Analyze your data.*

    As shown in Figure 7.3, use the pull-down menus to choose "Analyze", "Correlate", and "Bivariate...".

    Use the arrow button to place both variables with your data values in the box labeled "Variables:" as shown in Figure 7.4. Then, in the "Correlation

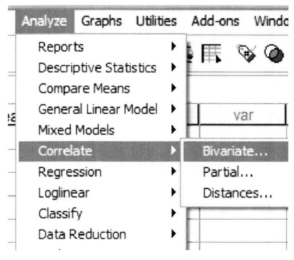

FIGURE 7.3

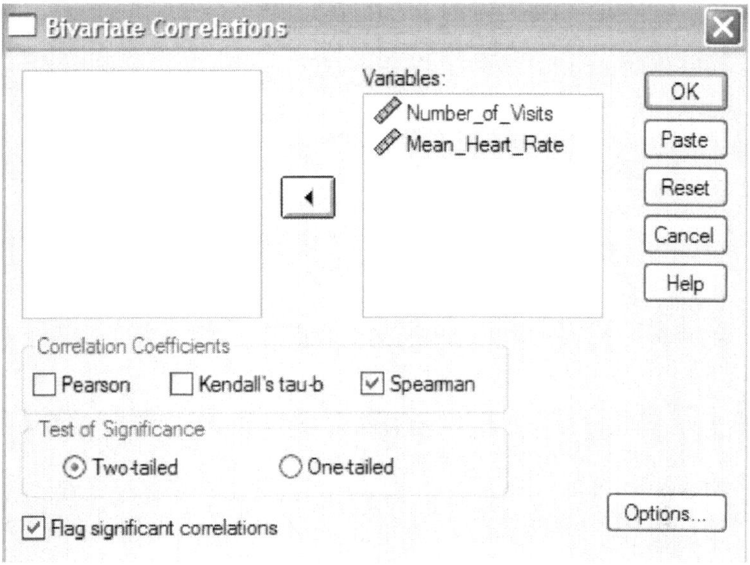

COMPUTING THE SPEARMAN RANK-ORDER CORRELATION COEFFICIENT

FIGURE 7.4

Coefficients" box, uncheck "Pearson" and check "Spearman". Finally, click "OK" to perform the analysis.

4. *Interpret the results from the SPSS Output window.*

Correlations

			Number_of_Visits	Mean_Heart_Rate
Spearman's rho	Number_of_Visits	Correlation Coefficient	1.000	-.860**
		Sig. (2-tailed)	.	.000
		N	13	13
	Mean_Heart_Rate	Correlation Coefficient	-.860**	1.000
		Sig. (2-tailed)	.000	.
		N	13	13

**. Correlation is significant at the 0.01 level (2-tailed).

The SPSS output table provides the Spearman rank-order correlation coefficient ($r_s = -0.860$) labeled Spearman's rho. It also returns the number of pairs ($n = 13$) and the two-tailed significance ($p \approx 0.000$). In this example, the significance is not actually zero. The reported value does not return enough digits to show the significance's actual precision.

Based on the results from SPSS, the Spearman rank-order correlation coefficient was significant ($r_{s(11)} = -0.860$, $p < 0.05$). Based on these data, we can state that there is a very strong inverse relationship between adult female resting heart rate and the frequency of visits to the gym.

7.5 COMPUTING THE POINT-BISERIAL AND BISERIAL CORRELATION COEFFICIENTS

The point-biserial correlation and biserial correlation are statistical procedures for use with dichotomous variables. A dichotomous variable is simply a measure of two conditions. A dichotomous variable is either discrete or continuous. A discrete dichotomous variable has no particular order and might include such examples as gender (male versus female) or a coin toss (heads versus tails). A continuous dichotomous variable has some type of order to the two conditions and might include measurements such as pass/fail or young/old. Finally, since the point-biserial and biserial correlations each involve an interval scale analysis, they are special cases of the Pearson product-moment correlation.

7.5.1 Correlation of a Dichotomous Variable and an Interval Scale Variable

The point-biserial correlation is a statistical procedure to measure the relationship between a discrete dichotomous variable and an interval scale variable. Use Formula 7.8 to determine the point-biserial correlation coefficient, r_{pb}.

$$r_{pb} = \frac{\bar{x}_p - \bar{x}_q}{s} \sqrt{P_p P_q} \qquad (7.8)$$

where $\bar{x}_p$ is the mean of the interval variable's values associated with the dichotomous variable's first category, $\bar{x}_q$ is the mean of the interval variable's values associated with the dichotomous variable's second category, s is the standard deviation of the variable on the interval scale, P_p is the proportion of the interval variable values associated with the dichotomous variable's first category, and P_q is the proportion of the interval variable values associated with the dichotomous variable's second category.

Recall the formulas for mean (7.9) and standard deviation (7.10) are below.

$$\bar{x} = \sum x_i / n \qquad (7.9)$$

and

$$s = \sqrt{\frac{\sum (x_i - \bar{x})^2}{n - 1}} \qquad (7.10)$$

where $\sum x_j$ is the sum of the values in the sample and n is the number of values in the sample.

The biserial correlation is a statistical procedure to measure the relationship between a continuous dichotomous variable and an interval scale variable. Use Formula 7.11 to determine the biserial correlation coefficient, r_b.

$$r_b = \left[\frac{\bar{x}_p - \bar{x}_q}{s_x} \right] \frac{P_p P_q}{y} \qquad (7.11)$$

COMPUTING THE POINT-BISERIAL AND BISERIAL CORRELATION COEFFICIENTS

where $\bar{x}_p$ is the mean of the interval variable's values associated with the dichotomous variable's first category, $\bar{x}_q$ is the mean of the interval variable's values associated with the dichotomous variable's second category, s_x is the standard deviation of the variable on the interval scale, P_p is the proportion of the interval variable values associated with the dichotomous variable's first category, P_q is the proportion of the interval variable values associated with the dichotomous variable's second category, and y is the height of the unit normal curve ordinate at the point dividing P_p and P_q (see Figure 7.5)

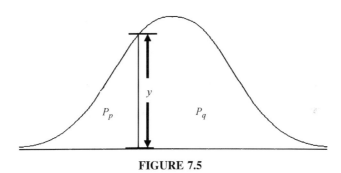

FIGURE 7.5

You may use Table B.1 or Formula 7.12 to find the height of the unit normal curve ordinate, y.

$$y = \frac{1}{\sqrt{2\pi}} e^{-z^2/2} \qquad (7.12)$$

where e is the natural log base and approximately equal to 2.718282 and z is the z-score at the point dividing P_p and P_q.

Formula 7.13 is the relationship between the point-biserial and the biserial correlation coefficients. This formula is necessary to find the biserial correlation coefficient because SPSS determines only the point-biserial correlation coefficient.

$$r_b = r_{pb} \frac{\sqrt{P_p P_q}}{y} \qquad (7.13)$$

After the correlation coefficient is determined, it must be examined for significance. Small samples allow one to reference a table of critical values, such as Table B.8. However, if the sample size, n, exceeds those available from the table, then a large sample approximation may be performed. For large samples, compute a z-score and use a table with the normal distribution (see Table B.1) to obtain a critical region of z-scores. As described earlier in this chapter, Formula 7.6 may be used to find the z-score of a correlation coefficient for large samples.

7.5.2 Correlation of a Dichotomous Variable and a Rank-Order Variable

As explained earlier, the above point-biserial and biserial correlation procedures involve a dichotomous variable and an interval scale variable. If the correlation was a dichotomous variable and a rank-order variable, a slightly different approach is needed.

To find the point-biserial correlation coefficient for a discrete dichotomous variable and a rank-order variable, simply use the Spearman rank-order described earlier and assign arbitrary values to the dichotomous variable, such as 0 and 1. To find the biserial correlation coefficient for a continuous dichotomous variable and a rank-order variable, use the same procedure and then apply Formula 7.13.

7.5.3 Sample Point-Biserial Correlation (Small Data Samples)

A researcher in a psychological lab investigated gender differences. She wished to compare male and female ability to recognize and remember visual details. She used 17 participants (8 males and 9 females) who were initially unaware of the actual experiment. First, she placed each one of them alone in a room with various objects and asked them to wait. After 10 min., she asked each of the participants to complete a 30-question posttest relating to several details in the room. Table 7.8 shows the participants' genders and posttest scores.

TABLE 7.8

Participant	Gender	Posttest Score
1	M	7
2	M	19
3	M	8
4	M	10
5	M	7
6	M	15
7	M	6
8	M	13
9	F	14
10	F	11
11	F	18
12	F	23
13	F	17
14	F	20
15	F	14
16	F	24
17	F	22

The researcher wishes to determine if a relationship exists between the two variables and the relative strength of the relationship. Gender is a discrete dichotomous variable and visual detail recognition is an interval scale variable. Therefore, we will use a point-biserial correlation.

1. *State the null and research hypotheses.*
   The null hypothesis, shown below, states that there is no correlation between gender and visual detail recognition. The research hypothesis states that there is a correlation between gender and visual detail recognition.
   The null hypothesis is

   H_0: $\rho_{pb} = 0$

   The research hypothesis is

   H_A: $\rho_{pb} \neq 0$

2. *Set the level of risk (or the level of significance) associated with the null hypothesis.*
   The level of risk, also called an alpha (α), is frequently set at 0.05. We will use an alpha of 0.05 in our example. In other words, there is a 95% chance that any observed statistical difference will be real and not due to chance.

3. *Choose the appropriate test statistic.*
   As stated earlier, we decided to analyze the relationship between the two variables. A correlation will provide the relative strength of the relationship between the two variables. Gender is a discrete dichotomous variable and visual detail recognition is an interval scale variable. Therefore, we will use a point-biserial correlation.

4. *Compute the test statistic.*
   First, compute the standard deviation of all values from the interval scale data. It is helpful to organize the data as shown in Table 7.9.

TABLE 7.9

Participant	Gender	Posttest Score	$x_i - \bar{x}$	$(x_i - \bar{x})^2$
1	M	7	−7.59	57.58
2	M	19	4.41	19.46
3	M	8	−6.59	43.40
4	M	10	−4.59	21.05
5	M	7	−7.59	57.58
6	M	15	0.41	0.17
7	M	6	−8.59	73.76
8	M	13	−1.59	2.52
9	F	14	−0.59	0.35
10	F	11	−3.59	12.88
11	F	18	3.41	11.64
12	F	23	8.41	70.76
13	F	17	2.41	5.82
14	F	20	5.41	29.29
15	F	14	−0.59	0.35
16	F	24	9.41	88.58
17	F	22	7.41	54.93
		$\sum x_i = 248$		$\sum (x_i - \bar{x})^2 = 550.12$

Using the summations from Table 7.9, calculate the mean and the standard deviation for the interval data.

$$\bar{x} = \sum x_i/n$$
$$= 248/17$$
$$= 14.59$$

$$s_x = \sqrt{\frac{\sum(x_i-\bar{x})^2}{n-1}} = \sqrt{\frac{550.12}{17-1}} = \sqrt{34.38}$$
$$= 5.86$$

Next, compute the means and proportions of the values associated with each item from the dichotomous variable. The mean males' posttest score was

$$\bar{x}_p = \sum x_p/n_M$$
$$= (7+19+8+10+7+15+6+13)/8$$
$$= 10.63$$

The mean females' posttest score was

$$\bar{x}_q = \sum x_q/n_F$$
$$= (14+11+18+23+17+20+14+24+22)/9$$
$$= 18.11$$

The males' proportion was

$$P_p = n_M/n$$
$$= 8/17$$
$$= 0.47$$

The females' proportion was

$$P_q = n_F/n$$
$$= 9/17$$
$$= 0.53$$

Now, compute the point-biserial correlation coefficient using the values computed above.

$$r_{pb} = \frac{\bar{x}_p - \bar{x}_q}{s_x}\sqrt{P_p P_q}$$
$$= \frac{10.63-18.11}{5.86}\sqrt{(0.47)(0.53)}$$
$$= \frac{-7.49}{5.86}\sqrt{0.25} = (-1.28)(0.50)$$
$$= -0.637$$

The sign of the correlation coefficient depends on the order we managed our dichotomous variable. Since that was arbitrary, the sign is irrelevant. Therefore, we use the absolute value of the point-biserial correlation coefficient.

$$r_{pb} = 0.637$$

5. *Determine the value needed for rejection of the null hypothesis using the appropriate table of critical values for the particular statistic.*

   Table B.8 lists critical values for the Pearson product-moment correlation coefficient. Using the table of critical values requires that the degrees of freedom be known. Since $df = n - 2$ and $n = 17$, then $df = 17 - 2$. Therefore, $df = 15$. Since we are conducting a two-tailed test and $\alpha = 0.05$, the critical value is 0.482.

6. *Compare the obtained value to the critical value.*

   The critical value for rejecting the null hypothesis is 0.482 and the obtained value is $|r_{pb}| = 0.637$. If the critical value is less than or equal to the obtained value, we must reject the null hypothesis. If instead, the critical value is greater than the obtained value, we must not reject the null hypothesis. Since the critical value is less than the absolute value of the obtained value, we reject the null hypothesis.

7. *Interpret the results.*

   We rejected the null hypothesis, suggesting that there is a significant and moderately strong correlation between gender and visual detail recognition.

8. *Reporting the results.*

   The reporting of results for the point-biserial correlation should include such information as the number of participants (n), two variables that are being correlated, correlation coefficient (r_{pb}), degrees of freedom (df), p-value's relation to α, and the mean values of each dichotomous variable.

   For this example, a researcher compared male and female ability to recognize and remember visual details. Eight males ($n_M = 8$) and nine females ($n_F = 9$) participated in the experiment. The researcher measured participants' visual detail recognition with a 30-question test requiring participants to recall details in a room they had occupied. A point-biserial correlation produced significant results ($r_{pb(15)} = 0.637$, $p < 0.05$). These data suggest that there is a strong relationship between gender and visual detail recognition. Moreover, the mean scores on the detail recognition test indicate that males ($\bar{x}_M = 10.63$) recalled fewer details, while females ($\bar{x}_F = 18.11$) recalled more details.

7.5.4 Performing the Point-Biserial Correlation Using SPSS

We will analyze the data from the previous example using SPSS.

1. *Define your variables.*

   First, click the "Variable View" tab at the bottom of your screen. Then, type the names of your variables in the "Name" column. As shown in Figure 7.6, the first variable is called "Gender" and the second variable is called "Posttest_Score".

COMPARING VARIABLES OF ORDINAL OR DICHOTOMOUS SCALES

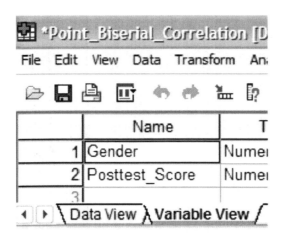

FIGURE 7.6

2. *Type in your values.*

   Click the "Data View" tab at the bottom of your screen as shown in Figure 7.7. Type in the values in the respective columns. Gender is a discrete dichotomous variable and SPSS needs a code to reference the values. We code male values with 0 and female values with 1. Any two values can be chosen for coding the data.

	Gender	Posttest_Score
1	.00	7.00
2	.00	19.00
3	.00	8.00
4	.00	10.00
5	.00	7.00
6	.00	15.00
7	.00	6.00
8	.00	13.00
9	1.00	14.00
10	1.00	11.00
11	1.00	18.00
12	1.00	23.00
13	1.00	17.00
14	1.00	20.00
15	1.00	14.00
16	1.00	24.00
17	1.00	22.00
18		

FIGURE 7.7

3. *Analyze your data.*

As shown in Figure 7.8, use the pull-down menus to choose "Analyze", "Correlate", and "Bivariate... ".

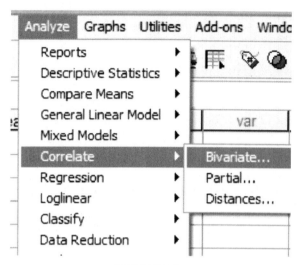

FIGURE 7.8

Use the arrow button near the middle of the window to place both variables with your data values in the box labeled "Variables:" as shown in Figure 7.9. In the "Correlation Coefficients" box, "Pearson" should remain checked since the Pearson product-moment correlation will perform an approximate point-biserial correlation. Finally, click "OK" to perform the analysis.

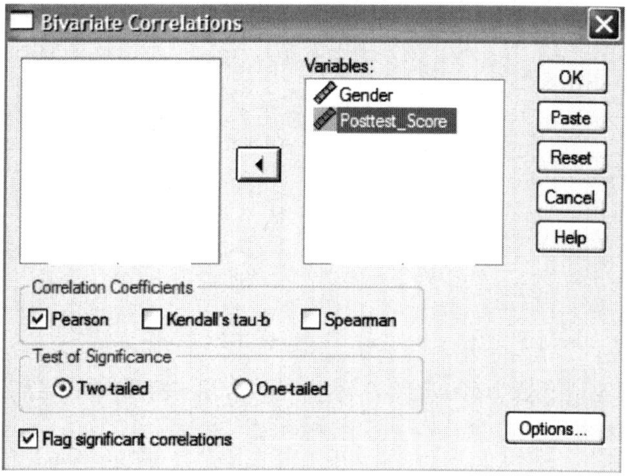

FIGURE 7.9

4. *Interpret the results from the SPSS Output window.*

Correlations

		Gender	Posttest_Score
Gender	Pearson Correlation	1	.657**
	Sig. (2-tailed)		.004
	N	17	17
Posttest_Score	Pearson Correlation	.657**	1
	Sig. (2-tailed)	.004	
	N	17	17

**. Correlation is significant at the 0.01 level (2-tailed).

The SPSS output table provides the Pearson product-moment correlation coefficient ($r = 0.657$). This correlation coefficient is approximately equal to the point-biserial correlation coefficient. It also returns the number of pairs ($n = 17$) and the two-tailed significance ($p = 0.004$).

Based on the results from SPSS, the point-biserial correlation coefficient was significant ($r_{pb(15)} = 0.657$, $p < 0.05$). Based on these data, we can state that there is a strong relationship between gender and visual detail recognition (as measured by the posttest).

7.5.5 Sample Point-Biserial Correlation (Large Data Samples)

A colleague of the researcher from the previous example wished to replicate the study investigating gender differences. As before, he compared male and female ability to recognize and remember visual details. He used 26 participants (14 males and 12 females) who were initially unaware of the actual experiment. Table 7.10 shows the participants' genders and posttest scores.

We will once again use a point-biserial correlation. However, we will use a large sample approximation to examine the results for significance since the sample size is large.

1. *State the null and research hypotheses.*

    The null hypothesis, shown below, states that there is no correlation between gender and visual detail recognition. The research hypothesis states that there is a correlation between gender and visual detail recognition.

    The null hypothesis is

    $H_0: \rho_{pb} = 0$

    The research hypothesis is

    $H_A: \rho_{pb} \neq 0$

2. *Set the level of risk (or the level of significance) associated with the null hypothesis.*

    The level of risk, also called an alpha (α), is frequently set at 0.05. We will use an alpha of 0.05 in our example. In other words, there is a 95% chance that any observed statistical difference will be real and not due to chance.

TABLE 7.10

Participant	Gender	Posttest Score
1	M	6
2	M	15
3	M	8
4	M	10
5	M	6
6	M	12
7	M	7
8	M	13
9	M	13
10	M	10
11	M	18
12	M	23
13	M	17
14	M	20
15	F	14
16	F	26
17	F	14
18	F	11
19	F	29
20	F	20
21	F	15
22	F	18
23	F	9
24	F	14
25	F	21
26	F	22

3. *Choose the appropriate test statistic.*

   As stated earlier, we decided to analyze the relationship between the two variables. A correlation will provide the relative strength of the relationship between the two variables. Gender is a discrete dichotomous variable and visual detail recognition is an interval scale variable. Therefore, we will use a point-biserial correlation.

4. *Compute the test statistic.*

   First, compute the standard deviation of all values from the interval scale data. Organize the data to manage the summations (see Table 7.11).

$$\bar{x} = \sum x_i/n$$
$$= 391/26$$
$$= 15.04$$

$$s_x = \sqrt{\frac{\sum(x_i - \bar{x})^2}{n-1}} = \sqrt{\frac{934.96}{26-1}} = \sqrt{37.40}$$
$$= 6.115$$

TABLE 7.11

Participant	Gender	Posttest Score	$x_i - \bar{x}$	$(x_i - \bar{x})^2$
1	M	6	−9.04	81.69
2	M	15	−0.04	0.00
3	M	8	−7.04	49.54
4	M	10	−5.04	25.39
5	M	6	−9.04	81.69
6	M	12	−3.04	9.23
7	M	7	−8.04	64.62
8	M	13	−2.04	4.16
9	M	13	−2.04	4.16
10	M	10	−5.04	25.39
11	M	18	2.96	8.77
12	M	23	7.96	63.39
13	M	17	1.96	3.85
14	M	20	4.96	24.62
15	F	14	−1.04	1.08
16	F	26	10.96	120.16
17	F	14	−1.04	1.08
18	F	11	−4.04	16.31
19	F	29	13.96	194.92
20	F	20	4.96	24.62
21	F	15	0.04	0.00
22	F	18	2.96	8.77
23	F	9	−6.04	36.46
24	F	14	−1.04	1.08
25	F	21	5.96	35.54
26	F	22	6.96	48.46
		$\sum x_i = 391$		$\sum (x_i - \bar{x})^2 = 934.96$

Next, compute the means and proportions of the values associated with each item from the dichotomous variable. The mean males' posttest score was

$$\bar{x}_p = \sum x_p / n_M$$
$$= (6 + 15 + 8 + 10 + 6 + 12 + 7 + 13 + 13 + 10 + 18 + 23 + 17 + 20)/14$$
$$= 12.71$$

The mean females' posttest score was

$$\bar{x}_q = \sum x_q / n_F$$
$$= (14 + 26 + 14 + 11 + 29 + 20 + 15 + 18 + 9 + 14 + 21 + 22)/12$$
$$= 17.75$$

The males' proportion was

$$P_p = n_M / n$$
$$= 14/26$$
$$= 0.54$$

The females' proportion was

$$P_q = n_F/n$$
$$= 12/26$$
$$= 0.46$$

Now, compute the point-biserial correlation coefficient using the values computed above.

$$r_{pb} = \frac{\bar{x}_p - \bar{x}_q}{s_x}\sqrt{P_p P_q} = \frac{12.71 - 17.75}{6.115}\sqrt{(0.54)(0.46)}$$

$$= \frac{-5.04}{6.115}\sqrt{0.25} = (-0.823)(0.50)$$

$$= -0.411$$

The sign of the correlation coefficient depends on the order we managed our dichotomous variable. Since that was arbitrary, the sign is irrelevant. Therefore, we use the absolute value of the point-biserial correlation coefficient.

$$r_{pb} = 0.411$$

Since our number of values is large, we will use a large sample approximation to examine the obtained value for significance. We will find a z-score for our data using an approximation to the normal distribution.

$$z^* = r_{pb}[\sqrt{n-1}]$$
$$= 0.411[\sqrt{26-1}]$$
$$= 2.055$$

5. *Determine the value needed for rejection of the null hypothesis using the appropriate table of critical values for the particular statistic.*

   Table B.1 is used to establish the critical region of z-scores. For a two-tailed test with $\alpha = 0.05$, we must not reject the null hypothesis if $-1.96 \leq z^* \leq 1.96$.

6. *Compare the obtained value to the critical value.*

   Notice that z^* is in the positive tail of the distribution ($2.055 > 1.96$). Therefore, we reject the null hypothesis. This suggests that the correlation between gender and visual detail recognition is real.

7. *Interpret the results.*

   We rejected the null hypothesis, suggesting that there is a significant and moderately weak correlation between gender and visual detail recognition.

8. *Reporting the results.*

   The reporting of results for the point-biserial correlation should include such information as the number of participants (n), two variables that are being correlated, correlation coefficient (r_{pb}), degrees of freedom (df), p-value's relation to α, and the mean values of each dichotomous variable.

   For this example, a researcher replicated a study that compared male and female ability to recognize and remember visual details. Fourteen males ($n_M = 14$) and 12 females ($n_F = 12$) participated in the experiment. The researcher measured participants' visual detail recognition with a 30-question test requiring participants to recall details in a room they had occupied. A point-biserial correlation produced significant results ($r_{pb(24)} = 0.411$, $p < 0.05$). These data suggest that there is a moderate relationship between gender and visual detail recognition. Moreover, the mean scores on the detail recognition test indicate that males ($\bar{x}_M = 12.71$) recalled fewer details, while females ($\bar{x}_F = 17.75$) recalled more details.

7.5.6 Sample Biserial Correlation (Small Data Samples)

A graduate anthropology department at a university wished to determine if its students' grade point averages (GPAs) can be used to predict performance on the department's comprehensive exam required for graduation. The comprehensive exam is graded on a pass/fail basis. Sixteen students participated in the comprehensive exam last year. Five of the students failed the exam. The GPAs and the exam performance of the students are displayed in Table 7.12.

Exam performance is a continuous dichotomous variable and GPA is an interval scale variable. Therefore, we will use a biserial correlation.

1. *State the null and research hypotheses.*

    The null hypothesis, shown below, states that there is no correlation between student GPA and comprehensive exam performance. The research hypothesis states that there is a correlation between student GPA and comprehensive exam performance.

    The null hypothesis is

    H_0: $\rho_b = 0$

TABLE 7.12

Participant	Exam Performance	GPA
1	F	3.5
2	F	3.4
3	F	3.3
4	F	3.2
5	F	3.6
6	P	4.0
7	P	3.6
8	P	4.0
9	P	4.0
10	P	3.8
11	P	3.9
12	P	3.9
13	P	4.0
14	P	3.8
15	P	3.5
16	P	3.6

COMPUTING THE POINT-BISERIAL AND BISERIAL CORRELATION COEFFICIENTS 147

The research hypothesis is

H_A: $\rho_b \neq 0$

2. *Set the level of risk (or the level of significance) associated with the null hypothesis.*

    The level of risk, also called an alpha (α), is frequently set at 0.05. We will use an alpha of 0.05 in our example. In other words, there is a 95% chance that any observed statistical difference will be real and not due to chance.

3. *Choose the appropriate test statistic.*

    As stated earlier, we decided to analyze the relationship between the two variables. A correlation will provide the relative strength of the relationship between the two variables. Exam performance is a continuous dichotomous variable and GPA is an interval scale variable. Therefore, we will use a biserial correlation.

4. *Compute the test statistic.*

    First, compute the standard deviation of all values from the interval scale data. Organize the data to manage the summations (see Table 7.13).

$$\bar{x} = \sum x_i / n$$
$$= 59.1/16$$
$$= 3.69$$

$$s_x = \sqrt{\frac{\sum (x_i - \bar{x})^2}{n-1}} = \sqrt{\frac{1.07}{16-1}} = \sqrt{0.071}$$
$$= 0.267$$

TABLE 7.13

Participant	Exam Performance	GPA	$x_i - \bar{x}$	$(x_i - \bar{x})^2$
1	F	3.5	−0.19	0.04
2	F	3.4	−0.29	0.09
3	F	3.3	−0.39	0.16
4	F	3.2	−0.49	0.24
5	F	3.6	−0.09	0.01
6	P	4.0	0.31	0.09
7	P	3.6	−0.09	0.01
8	P	4.0	0.31	0.09
9	P	4.0	0.31	0.09
10	P	3.8	0.11	0.01
11	P	3.9	0.21	0.04
12	P	3.9	0.21	0.04
13	P	4.0	0.31	0.09
14	P	3.8	0.11	0.01
15	P	3.5	−0.19	0.04
16	P	3.6	−0.09	0.01
		$\sum x_i = 59.1$		$\sum (x_i - \bar{x})^2 = 1.07$

Next, compute the means and proportions of the values associated with each item from the dichotomous variable. The mean GPA of the exam failures was

$$\bar{x}_p = \sum x_p/n_F$$
$$= (3.5+3.4+3.3+3.2+3.6)/5$$
$$= 3.4$$

The mean GPA of the exam passing scorers was

$$\bar{x}_q = \sum x_q/n_P$$
$$= (4.0+3.6+4.0+4.0+3.8+3.9+3.9+4.0+3.8+3.5+3.6)/11$$
$$= 3.8$$

The proportion of exam failures was

$$P_p = n_F/n$$
$$= 5/16$$
$$= 0.3125$$

The proportion of exam passing scorers was

$$P_q = n_P/n$$
$$= 11/16$$
$$= 0.6875$$

Now, determine the height of the unit normal curve ordinate, y, at the point dividing P_p and P_q. We could reference the table of values for the normal distribution, such as Table B.1, to find y. However, we will compute the value. Using Table B.1 also provides the z-score at the point dividing P_p and P_q, $z = 0.49$.

$$y = \frac{1}{\sqrt{2\pi}} e^{-z^2/2}$$
$$= \frac{1}{\sqrt{2\pi}} e^{-0.49^2/2} = \frac{1}{2.51} e^{-0.12} = (0.40)(0.89)$$
$$= 0.3538$$

Now, compute the biserial correlation coefficient using the values computed above.

$$r_b = \left[\frac{\bar{x}_p - \bar{x}_q}{s_x}\right] \frac{P_p P_q}{y}$$
$$= \left[\frac{3.4-3.8}{0.267}\right] \frac{(0.3125)(0.6875)}{0.3538} = (-1.60)(0.6072)$$
$$= -0.972$$

The sign of the correlation coefficient depends on the order we managed our dichotomous variable. A quick inspection of the variable means indicates that the GPA of the failures was smaller than the GPA of the passing scorers. Therefore, we should convert the biserial correlation coefficient to a positive value.

$$r_b = 0.972$$

5. *Determine the value needed for rejection of the null hypothesis using the appropriate table of critical values for the particular statistic.*

   Table B.8 lists critical values for the Pearson product-moment correlation coefficient. The table requires the degrees of freedom and $df = n - 2$. In this study, $n = 16$ and $df = 16 - 2$. Therefore, $df = 14$. Since we are conducting a two-tailed test and $\alpha = 0.05$, the critical value is 0.497.

6. *Compare the obtained value to the critical value.*

   The critical value for rejecting the null hypothesis is 0.497 and the obtained value is $|r_b| = 0.972$. If the critical value is less than or equal to the obtained value, we must reject the null hypothesis. If instead, the critical value is greater than the obtained value, we must not reject the null hypothesis. Since the critical value is less than the absolute value of the obtained value, we reject the null hypothesis.

7. *Interpret the results.*

   We rejected the null hypothesis, suggesting that there is a significant and very strong correlation between student GPA and comprehensive exam performance.

8. *Reporting the results.*

   The reporting of results for the biserial correlation should include such information as the number of participants (n), two variables that are being correlated, correlation coefficient (r_b), degrees of freedom (df), p-value's relation to α, and the mean values of each dichotomous variable.

   For this example, a researcher compared the GPAs of graduate anthropology students who passed their comprehensive exam with students who failed the exam. Five students failed the exam ($n_F = 5$) and 11 students passed it ($n_P = 11$). The researcher compared student GPA and comprehensive exam performance. A biserial correlation produced significant results ($r_{b(14)} = 0.972$, $p < 0.05$). The data suggest that there is an especially strong relationship between student GPA and comprehensive exam performance. Moreover, the mean GPA of the failing students ($\bar{x}_{failure} = 3.4$) and passing students ($\bar{x}_{passing} = 3.8$) indicates that the relationship is a direct correlation.

7.5.7 Performing the Biserial Correlation Using SPSS

SPSS does not compute the biserial correlation coefficient. To do so, Field (2005) has suggested using SPSS to perform a Pearson product-moment correlation (as described above) and then applying Formula 7.13. However, this procedure will only produce an approximation of the biserial correlation coefficient and we recommend you use a spreadsheet with the procedure we described for the biserial correlation.

7.6 EXAMPLES FROM THE LITERATURE

Below are varied examples of the nonparametric procedures described in this chapter. We have summarized each study's research problem and researchers' rationale(s) for choosing a nonparametric approach. We encourage you to obtain these studies if you are interested in their results.

- Greiner, C. S., & Smith, B. (2006). Determining the effect of selected variables on teacher retention. *Education, 126*(4), 653–659.

    Greiner and Smith investigated factors that might affect teacher retention. When they examined the relationship between the Texas state-mandated teacher certification examination and teacher retention, they used a point-biserial correlation. The researchers used the point-biserial since teacher retention was measured as a discrete dichotomous variable.

- Blumberg, F. C., & Sokol, L. M. (2004). Boys' and girls' use of cognitive strategy when learning to play video games. *The Journal of General Psychology 131*(2), 151–158.

    Blumberg and Sokol examined gender differences in the cognitive strategies that second- and fifth-grade children use when they learn how to play a video game. In part of the study, participants were classified as frequent players or infrequent players. That classification was correlated with game performance. Since player frequency was a discrete dichotomy, the researchers chose a point-biserial correlation.

- McMillian, J., Morgan, S. A., & Ament, P. (2006). Acceptance of male registered nurses by female registered nurses. *Journal of Nursing Scholarship, 38*(1), 100–106.

    McMillian, Morgan, and Ament investigated the attitudes of female registered nurses toward male registered nurses. The researchers performed several analyses with a variety of statistical tests. In one analysis, they used a Spearman rank-order correlation to examine the relationship between town population and the participants' responses on an attitude inventory. The attitude inventory was a modified instrument to measure level of sexist attitude. Participants indicated agreement or disagreement with statements using a four-point scale. The Spearman rank-order correlation was chosen because the attitude inventory resembled an ordinal scale.

- Fitzgerald, L. M., Delitto, A., & Irrgang, J. J. (2007). Validation of the clinical internship evaluation tool. *Physical Therapy, 87*(7), 844–860.

    Fitzgerald, Delitto, and Irrgang examined the validity of an instrument designed to measure the performance of physical therapy interns. They used a correlation analysis to examine the relationship between two measures of clinical competence. Since one of the measures was ordinal, the researchers used a Spearman rank-order correlation.

- Flannelly, K. J., Strock, A. L., & Weaver, A. J. (2005). A review of research on the effects of religion on adolescent tobacco use published between 1990 and 2003. *Adolescence, 40*, 761–776.

Flannelly, Strock, and Weaver reviewed the research literature of studies that investigated the effects of religion on adolescent tobacco use. The authors used a biserial correlation to compare studies' effect (no effect versus effect) with sample size.

7.7 SUMMARY

The relationship between two variables can be compared with a correlation analysis. If any of the variables are ordinal or dichotomous, a nonparametric correlation is useful. The Spearman rank-order correlation, also called the Spearman's rho, is used to compare the relationship involving ordinal, or rank-ordered, variables. The point-biserial and biserial correlations are used to compare the relationship between two variables if one of the variables is dichotomous. The parametric equivalent to these correlations is the Pearson product-moment correlation.

In this chapter, we described how to perform and interpret Spearman rank-order, point-biserial, and biserial correlations. We also explained how to perform the procedures using SPSS. Finally, we offered varied examples of these nonparametric statistics from the literature. The next chapter will involve comparing nominal scale data.

7.8 PRACTICE QUESTIONS

1. The business department at a small college wanted to compare the relative class rank of its MBA graduates with their fifth-year salaries. The data collected by the department are presented in Table 7.14. Compare the graduates' class rank with their fifth-year salaries.

TABLE 7.14

Relative Class Rank	Fifth-Year Salary
1	$83,450
2	$67,900
3	$89,000
4	$80,500
5	$91,000
6	$55,440
7	$101,300
8	$50,560
9	$76,050

Use a two-tailed Spearman rank-order correlation with $\alpha = 0.05$ to determine if a relationship exists between the two variables. Report your findings.

2. A researcher was contracted by the military to assess soldiers' perception of a new training program's effectiveness. Fifteen soldiers serving in Iraq participated in the program. The researcher used a survey to measure the soldiers' perceptions of the program's effectiveness. The survey used a Likert-type scale that ranged from 5 = strongly agree to 1 = strongly disagree. Using the data presented in Table 7.15, compare the soldiers' average survey scores with the total number of years the soldiers had been serving.

TABLE 7.15

Average Survey Score	Years of Service
4.0	18
4.0	15
2.4	2
4.2	13
3.4	4
4.0	10
5.0	24
1.8	4
3.2	9
2.5	5
2.5	3
3.0	8
3.6	16
4.6	14
4.8	12

Use a two-tailed Spearman rank-order correlation with $\alpha = 0.05$ to determine if a relationship exists between the two variables. Report your findings.

3. A middle school history teacher wished to determine if there is a connection between gender and history knowledge among eighth-grade gifted students. The teacher administered a 50-item test at the beginning of the school year to 16 gifted eighth-grade students. The scores from the test are presented in Table 7.16.

Use a two-tailed point-biserial correlation with $\alpha = 0.05$ to determine if a relationship exists between the two variables. Report your findings.

4. A researcher wished to determine if there is a connection between poverty and self-esteem. Income level was used to classify 18 participants as either below poverty or above poverty. Participants completed a 20-item survey to measure self-esteem. The scores from the survey are reported in Table 7.17.

Use a two-tailed biserial correlation with $\alpha = 0.05$ to determine if a relationship exists between the two variables. Report your findings.

TABLE 7.16

Participant	Gender	Posttest Score
1	M	44
2	M	30
3	M	50
4	M	33
5	M	37
6	M	35
7	M	36
8	F	29
9	F	39
10	F	33
11	F	50
12	F	45
13	F	37
14	F	30
15	F	34
16	F	50

TABLE 7.17

Participant	Poverty Level	Survey Score
1	Above	15
2	Above	19
3	Above	15
4	Above	20
5	Above	7
6	Above	12
7	Above	3
8	Above	15
9	Below	9
10	Below	5
11	Below	13
12	Below	13
13	Below	11
14	Below	10
15	Below	8
16	Below	9
17	Below	10
18	Below	17

7.9 SOLUTIONS TO PRACTICE QUESTIONS

1. The results from the analysis are displayed in the SPSS Output below.

Correlations

			Class_Rank	Fifth_Yr_Salary
Spearman's rho	Class_Rank	Correlation Coefficient	1.000	-.217
		Sig. (2-tailed)	.	.576
		N	9	9
	Fifth_Yr_Salary	Correlation Coefficient	-.217	1.000
		Sig. (2-tailed)	.576	.
		N	9	9

The results from the Spearman rank-order correlation ($r_s = -0.217$, $p > 0.05$) did not produce significant results. Based on these data, we can state that there is no clear relationship between graduates' relative class rank and fifth-year salary.

2. The results from the analysis are displayed in the SPSS Output below.

Correlations

			Survey_Score	Years_of_Service
Spearman's rho	Survey_Score	Correlation Coefficient	1.000	.806**
		Sig. (2-tailed)	.	.000
		N	15	15
	Years_of_Service	Correlation Coefficient	.806**	1.000
		Sig. (2-tailed)	.000	.
		N	15	15

**. Correlation is significant at the 0.01 level (2-tailed).

The results from the Spearman rank-order correlation ($r_s = 0.806$, $p < 0.05$) produced significant results. Based on these data, we can state that there is a very strong correlation between soldiers' survey scores concerning the new program's effectiveness and their total years of military service.

3. The results from the point-biserial correlation ($r_{pb} = 0.047$, $p > 0.05$) did not produce significant results. Based on these data, we can state that there is no clear relationship between eight-grade gifted students' gender and their score on the history knowledge test administered by the teacher.

    Note: The results obtained from using SPSS are $r_{pb} = 0.049$, $p > 0.05$.

4. The results from the biserial correlation ($r_b = 0.372$, $p > 0.05$) did not produce significant results. Based on these data, we can state that there is no clear relationship between poverty level and self-esteem.

8

TESTS FOR NOMINAL SCALE DATA: CHI-SQUARE AND FISHER EXACT TEST

8.1 OBJECTIVES

In this chapter, you will learn the following items.

- How to perform a chi-square goodness-of-fit test.
- How to perform a chi-square goodness-of-fit test using SPSS.
- How to perform chi-square test for independence.
- How to perform chi-square test for independence using SPSS.
- How to perform the Fisher exact test.
- How to perform Fisher exact test using SPSS.

8.2 INTRODUCTION

Sometimes, data are best collected or conveyed nominally, or categorically. These data are represented by counting the number of times a particular event or condition occurs. In such cases, you may be seeking to determine if a given set of counts, or frequencies, statistically matches some known, or expected, set. Or, you may wish to determine if two or more categories are statistically independent. In either case, we can use a nonparametric procedure to analyze nominal data.

Nonparametric Statistics for Non-Statisticians, Gregory W. Corder and Dale I. Foreman
Copyright © 2009 John Wiley & Sons, Inc.

In this chapter, we present three procedures for examining nominal data: chi-square (χ^2) goodness-of-fit, chi-square test for independence, and the Fisher exact test. We will also explain how to perform the procedures using SPSS. Finally, we offer varied examples of these nonparametric statistics from the literature.

8.3 THE CHI-SQUARE GOODNESS-OF-FIT TEST

Some situations in research involve investigations and questions about relative frequencies and proportions for a distribution. Some examples might include a comparison of the number of women pediatricians to the number of men pediatricians, a search for significant changes in the proportion of students entering a writing contest over 5 years, or an analysis of customer preference of three candy bar choices. Each of these examples asks a question about a proportion in the population.

When comparing proportions, we are not measuring a numerical score for each individual. Instead, we classify each individual into a category. We then find out what proportion of the population is classified into each category. The chi-square goodness-of-fit test is designed to answer this type of question.

The chi-square goodness-of-fit test uses sample data to test the hypothesis about the proportions of the population distribution. The test determines how well the sample proportions fit the proportions specified in the null hypothesis.

8.3.1 Computing the Chi-Square Goodness-of-Fit Test Statistic

The chi-square goodness-of-fit test is used to determine how well the obtained sample proportions or frequencies for a distribution fit the population proportions or frequencies specified in the null hypothesis. The chi-square statistic can be used when two or more categories are involved in the comparison. Formula 8.1 is referred to as Pearson's chi-square and is used to determine the χ^2 statistic.

$$\chi^2 = \sum \frac{(f_o - f_e)^2}{f_e} \qquad (8.1)$$

where f_o is the observed frequency (the data) and f_e is the expected frequency (the hypothesis). Use Formula 8.2 to determine the expected frequency, f_e.

$$f_e = P_i n \qquad (8.2)$$

where P_i is a category's frequency proportion with respect to the other categories and n is the sample size of all categories and $\sum f_o = n$.

Use Formula 8.3 to determine the degrees of freedom for the chi-square test.

$$df = C - 1 \qquad (8.3)$$

where C is the number of categories.

THE CHI-SQUARE GOODNESS-OF-FIT TEST

8.3.2 Sample Chi-Square Goodness-of-Fit Test (Category Frequencies Equal)

A marketing firm is conducting a study to determine if there is a significant preference for a type of chicken that is served as a fast food. The target group is college students. It is assumed that there is no preference when the study is started. The types of food that are being compared are chicken sandwich, chicken strips, chicken nuggets, and chicken taco.

The sample size for this study was $n = 60$. The data in Table 8.1 represent the observed frequencies for the 60 participants who were surveyed at the fast food restaurants.

TABLE 8.1

Chicken Sandwich	Chicken Strips	Chicken Nuggets	Chicken Taco
10	25	18	7

We want to determine if there is any preference for one of the four chicken fast foods that were purchased to eat by the college students. Since the data only need to be classified into categories, and no sample mean or sum of squares needs to be calculated, the chi-square goodness-of-fit test can be used to test the nonparametric data.

1. *State the null and research hypotheses.*

    The null hypothesis, shown below, states that there is no preference among the different categories. There is an equal proportion or frequency of participants selecting each type of fast food that uses chicken. The research hypothesis states that one or more of the chicken fast foods is preferred over the others by the college student.

    The null hypothesis is

    H_0: In the population of college students, there is no preference of one chicken fast food over any other. Thus, the four fast food types are selected equally often and the population distribution has the proportions shown in Table 8.2.

TABLE 8.2

Chicken Sandwich	Chicken Strips	Chicken Nuggets	Chicken Taco
25%	25%	25%	25%

H_A: In the population of college students, there is at least one chicken fast food preferred over the others.

2. *Set the level of risk (or the level of significance) associated with the null hypothesis.*
   The level of risk, also called an alpha (α), is frequently set at 0.05. We will use an alpha of 0.05 in our example. In other words, there is a 95% chance that any observed statistical difference will be real and not due to chance.
3. *Choose the appropriate test statistic.*
   The data are obtained from 60 college students who eat fast food chicken. Each student was asked which of the four chicken types of food he or she purchased to eat and the result was tallied under the corresponding category type. The final data consisted of frequencies for each of the four types of chicken fast foods. These categorical data, which are represented by frequencies or proportions, are analyzed using the chi-square goodness-of-fit test.
4. *Compute the test statistic.*
   First, tally the observed frequencies, f_o, for the 60 students who were in the study. Use these data to create the observed frequency table shown in Table 8.3.

TABLE 8.3

	Chicken Sandwich	Chicken Strips	Chicken Nuggets	Chicken Taco
Observed frequencies	10	25	18	7

Next, calculate the expected frequency for each category. In this case, the expected frequency, f_e, will be the same for all four categories, since our research problem assumes that all categories are equal.

$$f_e = P_i n = \frac{1}{4}(60)$$

$$f_e = 15$$

Table 8.4 presents the expected frequencies for each category.

Using the values for the observed and expected frequencies, the chi-square statistic may be calculated.

TABLE 8.4

	Chicken Sandwich	Chicken Strips	Chicken Nuggets	Chicken Taco
Expected frequencies	15	15	15	15

$$\chi^2 = \sum \frac{(f_o - f_e)^2}{f_e}$$
$$= \frac{(10-15)^2}{15} + \frac{(25-15)^2}{15} + \frac{(18-15)^2}{15} + \frac{(7-15)^2}{15}$$

$$= \frac{(-5)^2}{15} + \frac{(10)^2}{15} + \frac{(3)^2}{15} + \frac{(-8)^2}{15}$$

$$= \frac{25}{15} + \frac{100}{15} + \frac{9}{15} + \frac{64}{15}$$

$$= 1.67 + 6.67 + 0.60 + 4.27$$

$$= 13.21$$

5. *Determine the value needed for rejection of the null hypothesis using the appropriate table of critical values for the particular statistic.*

    Before we go to the table of critical values, we must determine the degrees of freedom, df. In this example, there are four categories, $C = 4$. To find the degrees of freedom, use df $= C - 1 = 4 - 1$. Therefore, df $= 3$.

    Now, we use Table B.2, which lists the critical values for the chi-square. The critical value is found in the chi-square table for three degrees of freedom, df $= 3$. Since we set $\alpha = 0.05$, the critical value is 7.81. A calculated value that is greater than or equal to 7.81 will lead us to reject the null hypothesis.

6. *Compare the obtained value to the critical value.*

    The critical value for rejecting the null hypothesis is 7.81 and the obtained value is $\chi^2 = 13.21$. If the critical value is less than or equal to the obtained value, we must reject the null hypothesis. If instead, the critical value exceeds the obtained value, we do not reject the null hypothesis. Since the critical value is less than our obtained value, we must reject the null hypothesis.

    Note: The critical value for $\alpha = 0.01$ is 11.34. Since the obtained value is 13.21, a value greater than 11.34, the data indicate that the results are highly significant.

7. *Interpret the results.*

    We rejected the null hypothesis, suggesting that there is a real difference among chicken fast food choices preferred by college students. In particular, the data show that a larger portion of the students preferred the chicken strips and only a few of them preferred the chicken taco.

8. *Reporting the results.*

    The reporting of results for the chi-square goodness-of-fit test should include such information as the total number of participants in the sample and the number that were classified in each category. In some cases, bar graphs are good methods of presenting the data. In addition, include the χ^2 statistic, degrees of freedom, and p-value's relation to α. For this study, the number of students who ate each type of chicken fast food should be either noted in a table or graphed on a bar graph. The probability, $p < 0.01$, should also be indicated with the data to show the degree of significance of the chi-square.

    For this example, 60 college students were surveyed to determine which fast food type of chicken they purchased to eat. The four choices were chicken sandwich, chicken strips, chicken nuggets, and chicken taco. Student choices were 10, 25, 18, and 7, respectively. The chi-square goodness-of-fit test was significant ($\chi^2_{(3)} = 13.21, p < 0.01$). Based on these results, a larger portion of the students preferred the chicken strips, while only a few students preferred the chicken taco.

8.3.3 Sample Chi-Square Goodness-of-Fit Test (Category Frequencies Not Equal)

Sometimes research is being conducted in an area where there is a basis for different expected frequencies in each category. In this case, the null hypothesis will indicate different frequencies for each of the categories according to the expected values. These values are usually obtained from previous data that were collected in similar studies.

In this study, a school system uses three different physical fitness programs because of scheduling requirements. A researcher is studying the effect of the programs on tenth-grade students' one-mile run performance. Three different physical fitness programs were used by the school system and are described below.

- *Program 1*: delivers health and physical education in 9-week segments with an alternating rotation of 9 straight weeks of health education and then 9 straight weeks of physical education.
- *Program 2*: delivers health and physical education everyday with 30 minutes for health, 10 minutes for dress-out time, and 50 minutes of actual physical activity.
- *Program 3*: delivers health and physical education in 1-week segments with an alternating rotation of 1 week of health education and then 1 week of physical education.

Using students who participated in all three programs, the researcher is comparing these programs based on student performances on the one-mile run. The researcher recorded the program in which each student received the most benefit. Two hundred fifty students had participated in all three programs. The results for all of the students are recorded in Table 8.5.

TABLE 8.5

Program 1	Program 2	Program 3
110	55	85

We want to determine if the above frequency distribution is different from previous studies. Since the data only need to be classified into categories, and no sample mean or sum of squares needs to be calculated, the chi-square goodness-of-fit test can be used to test the nonparametric data.

1. *State the null and research hypotheses.*

    The null hypothesis states the proportion of students who benefited most from one of the three programs based upon a previous study. As shown in Table 8.6, there are unequal expected frequencies for the null hypothesis. The research hypothesis states that there is at least one of the three categories that

THE CHI-SQUARE GOODNESS-OF-FIT TEST

will have a different proportion or frequency from those identified in the null hypothesis.

The null hypothesis is

H_0: The proportions do not differ from the previously determined proportions shown in Table 8.6.

TABLE 8.6

Program 1	Program 2	Program 3
32%	22%	45%

H_A: The population distribution has a different shape from that specified in the null hypothesis.

2. *Set the level of risk (or the level of significance) associated with the null hypothesis.*

The level of risk, also called an alpha (α), is frequently set at 0.05. We will use an alpha of 0.05 in our example. In other words, there is a 95% chance that any observed statistical difference will be real and not due to chance.

3. *Choose the appropriate test statistic.*

The data are being obtained from the one-mile run performance of 250 tenth-grade students who participated in a school system's three health and physical education programs. Each student was categorized based on the plan in which he or she benefited most. The final data consisted of frequencies for each of the three plans. These categorical data, which are represented by frequencies or proportions, are analyzed using the chi-square goodness-of-fit test.

4. *Compute the test statistic.*

First, tally the observed frequencies for the 250 students who were in the study. This was performed by the researcher. Use the data to create the observed frequency table shown in Table 8.7.

TABLE 8.7

	Program 1	Program 2	Program 3
Observed frequencies	110	55	85

Next, calculate the expected frequencies for each category. In this case, the expected frequency will be different for each category. Each one will be based on proportions stated in the null hypothesis.

$$f_{ei} = P_i n$$

f_e for Program 1 = 0.32(250) = 80
f_e for Program 2 = 0.22(250) = 55
f_e for Program 3 = 0.46(250) = 115

Table 8.8 presents the expected frequencies for each category.

TABLE 8.8

	Program 1	Program 2	Program 3
Expected frequencies	80	55	115

Use the values for the observed and expected frequencies calculated above to calculate the chi-square statistic.

$$\chi^2 = \sum \frac{(f_o - f_e)^2}{f_e}$$

$$= \frac{(110-80)^2}{80} + \frac{(55-55)^2}{55} + \frac{(85-115)^2}{115}$$

$$= \frac{30^2}{80} + \frac{0^2}{55} + \frac{-30^2}{115}$$

$$= 11.25 + 0 + 7.83$$

$$= 19.08$$

5. *Determine the value needed for rejection of the null hypothesis using the appropriate table of critical values for the particular statistic.*

    Before we go to the table of critical values, we must determine the degrees of freedom, df. In this example, there are three categories, $C = 3$. To find the degrees of freedom, use $df = C - 1 = 3 - 1$. Therefore, $df = 2$.

    Now, we use Table B.2, which lists the critical values for the chi-square. The critical value is found in the chi-square table for two degrees of freedom, $df = 2$. Since we set $\alpha = 0.05$, the critical value is 5.99. A calculated value that is greater than 5.99 will lead us to reject the null hypothesis.

6. *Compare the obtained value to the critical value.*

    The critical value for rejecting the null hypothesis is 5.99 and the obtained value is $\chi^2 = 19.08$. If the critical value is less than or equal to the obtained value, we must reject the null hypothesis. If instead, the critical value exceeds the obtained value, we do not reject the null hypothesis. Since the critical value is less than our obtained value, we must reject the null hypothesis.

    Note: The critical value for $\alpha = 0.01$ is 9.21. Since the obtained value is 19.08, which is greater than the critical value, the data indicate that the results are highly significant.

7. *Interpret the results.*

    We rejected the null hypothesis, suggesting that there is a real difference in how the health and physical education program affects the performance of students on the one-mile run as compared to existing research. By comparing the expected frequencies of the past study and those obtained in the current study, it can be noted that the results from Program 2 did not change. Program 2 was least effective in both cases, with no difference in the two. Program 1 became more effective and Program 3 became less effective.

8. *Reporting the results.*

The reporting of results for the chi-square goodness-of-fit should include such information as the total number of participants in the sample, the number that were classified in each category, and the expected frequencies that are being used for comparison. It is also important to cite a source for the expected frequencies so that the decisions made from the study can be supported. In addition, include the χ^2 statistic, degrees of freedom, and p-value's relation to α. It is often a good idea to present a bar graph to display the observed and expected frequencies from the study. For this study, the probability, $p < 0.01$, should also be indicated with the data to show the degree of significance of the chi-square.

For this example, 250 tenth-grade students participated in three different health and physical education programs. Using one-mile run performance, students' program of greatest benefit was compared to results from past research. The chi-square goodness-of-fit test was significant ($\chi^2_{(2)} = 19.08$, $p < 0.01$). Based on these results, Program 2 was least effective in both cases, with no difference in the two. Program 1 became more effective and Program 3 became less effective.

8.3.4 Performing the Chi-Square Goodness-of-Fit Test Using SPSS

We will analyze the data from the above example using SPSS.

1. *Define your variables.*

First, click the "Variable View" tab at the bottom of your screen (see Figure 8.1). The chi-square goodness-of-fit test requires two variables: one variable to identify the categories and a second variable to identify the observed frequencies. Type the names of these variables in the "Name" column. In our example, we define the variables as "Program" and "count".

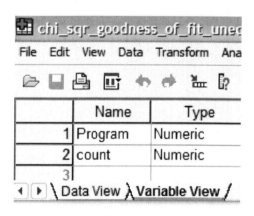

FIGURE 8.1

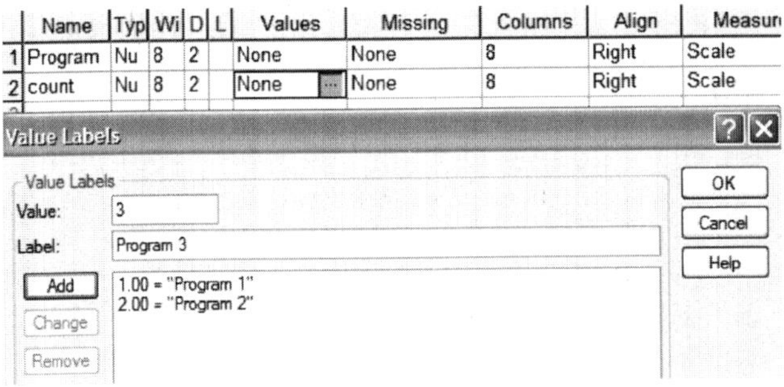

TESTS FOR NOMINAL SCALE DATA: CHI-SQUARE AND FISHER EXACT TEST

FIGURE 8.2

You must assign values to serve as a reference for each category in the observed frequencies variable. It is often easiest to assign each category a whole number value. As shown in Figure 8.2, our categories are "Program 1", "Program 2", and "Program 3". First, we selected the "count" variable and clicked the gray square in the "Values" field. Then, we set a value of 1 to equal "Program 1" and a value of 2 to equal "Program 2". As soon as we click the "Add" button, we will have set "Program 3" equal to 3. Repeat this procedure for the "Program" variable so that the output tables will display these labels.

2. *Type in your values.*

   Click the "Data View" tab at the bottom of your screen. First, enter the data for each category using the whole numbers you assigned to represent the categories. As shown in Figure 8.3, we entered the values "1", "2", and "3" in

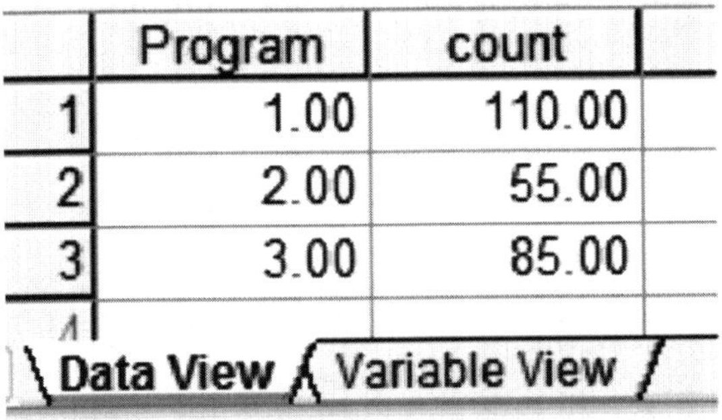

	Program	count
1	1.00	110.00
2	2.00	55.00
3	3.00	85.00

FIGURE 8.3

the "Program" variable. Second, enter the observed frequencies next to the corresponding category values. In our example, we entered the observed frequencies "110", "55", and "85".

3. *Analyze your data.*

   First, use the "Weight Cases" command to allow the observed frequencies variable to reference the categories variable. As shown in Figure 8.4, use the pull-down menus to choose "Data" and "Weight Cases…".

THE CHI-SQUARE GOODNESS-OF-FIT TEST

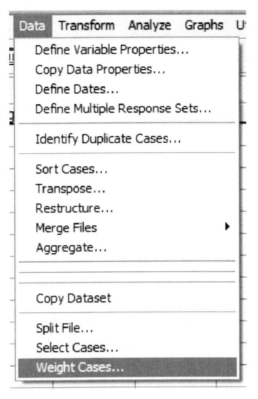

FIGURE 8.4

The default setting is "Do not weight cases". Click the circle next to "Weight cases by" as shown in Figure 8.5. Select the variable with the observed frequencies. Move that variable to the "Frequency Variable:" box by clicking the small arrow button. In our example, we have moved the "count" variable. Finally, click "OK".

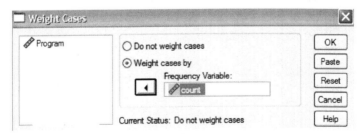

FIGURE 8.5

As shown in Figure 8.6, use the pull-down menus to choose "Analyze", "Nonparametric Tests", and "Chi-Square...".

166 TESTS FOR NOMINAL SCALE DATA: CHI-SQUARE AND FISHER EXACT TEST

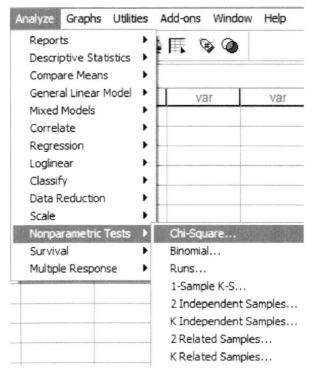

FIGURE 8.6

First, move the categories variable to the "Test Variable List:" box by selecting that variable and clicking the small arrow button near the center of the window. As shown in Figure 8.7, we have chosen the "Program" variable. Then, enter your

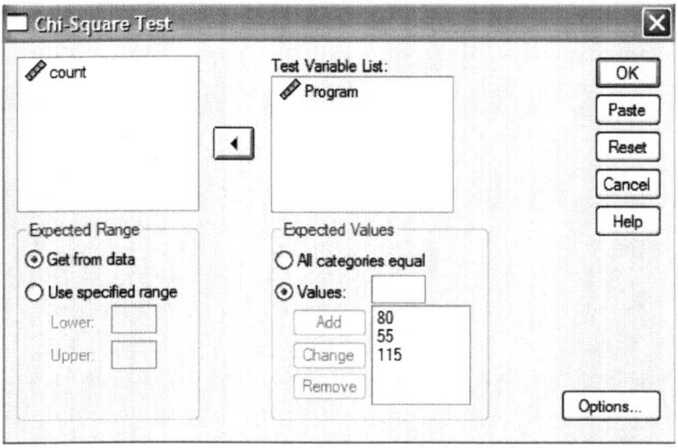

FIGURE 8.7

THE CHI-SQUARE TEST FOR INDEPENDENCE

"Expected Values". Notice that the option "All categories equal" is the default setting. Since this example does not have equal categories, we must select the "Values:" option to set the expected values. Enter the expected frequencies for each category in the order that they are listed in the "Data View". After you type in an expected frequency, click "Add". For our example, we have entered 80, 55, and 115, respectively. Finally, click "OK" to perform the analysis.

4. *Interpret the results from the SPSS Output window.*

Program

	Observed N	Expected N	Residual
Program 1	110	80.0	30.0
Program 2	55	55.0	.0
Program 3	85	115.0	-30.0
Total	250		

The SPSS output table above provides the observed and expected frequencies for each category and the total count.

Test Statistics

	Program
Chi-Square[a]	19.076
df	2
Asymp. Sig.	.000

a. 0 cells (.0%) have expected frequencies less than 5. The minimum expected cell frequency is 55.0.

The second SPSS output table provides the chi-square statistic ($\chi^2 = 19.076$), the degrees of freedom (df = 2), and the significance ($p \approx 0.000$).

Based on the results from SPSS, three programs were compared with unequal expected frequencies. The chi-square goodness-of-fit test was significant ($\chi^2_{(2)} = 19.08$, $p < 0.01$). Based on these results, Program 2 was least effective in both cases, with no difference in the two. Program 1 became more effective and Program 3 became less effective.

8.4 THE CHI-SQUARE TEST FOR INDEPENDENCE

Some research involves investigations of frequencies of statistical associations of two categorical attributes. Examples might include a sample of men and women who bought a pair of shoes or a shirt. The first attribute, A, is the gender of the shopper with two possible categories:

$$\text{men} = A_1$$
$$\text{women} = A_2$$

The second attribute, B, is the clothing type purchased by each individual:

$$\text{pair of shoes} = B_1$$
$$\text{shirt} = B_2$$

We will assume that each person purchased only one item, either a pair of shoes or a shirt. The entire set of data is then arranged into a joint-frequency distribution table. Each individual is classified into one category that is identified by a pair of categorical attributes (see Table 8.9).

TABLE 8.9

	A_1	A_2
B_1	(A_1, B_1)	(A_2, B_1)
B_2	(A_1, B_2)	(A_2, B_2)

The chi-square test for independence uses sample data to test the hypothesis that there is no statistical association between two categories; in this case, whether there is a significant association between the gender of the purchaser and the type of clothing purchased. The test determines how well the sample proportions fit the proportions specified in the null hypothesis.

8.4.1 Computing the Chi-Square Test for Independence

The chi-square test for independence is used to determine whether there is a statistical association between two categorical attributes. The chi-square statistic can be used when two or more categories are involved for two attributes. Formula 8.4 is referred to as Pearson's chi-square and is used to determine the χ^2 statistic.

$$\chi^2 = \sum_j \sum_k \frac{(f_{ojk} - f_{ejk})^2}{f_{ejk}} \tag{8.4}$$

where f_{ojk} is the observed frequency for cell $A_j B_k$ and f_{ejk} is the expected frequency for cell $A_j B_k$.

In tests for independence, the expected frequency, f_{ejk}, in any cell is found by multiplying the row total and the column total and dividing the product by the grand total, N. Use Formula 8.5 to determine the expected frequency, f_{ejk}.

$$f_{ejk} = \frac{(\text{freq } A_j)(\text{freq } B_k)}{N} \tag{8.5}$$

THE CHI-SQUARE TEST FOR INDEPENDENCE

The degrees of freedom, df, for the chi-square are found using Formula 8.6.

$$\mathrm{df} = (R-1)(C-1) \tag{8.6}$$

where R is the number of rows and C is the number of columns.

It is important to note that Pearson's chi-square formula returns a value that is too small when data form a 2 × 2 contingency table. This increases the chance of a Type I error. In such a circumstance, one might use the Yates's continuity correction shown in Formula 8.7.

$$\chi^2 = \sum_j \sum_k \frac{(f_{ojk} - f_{ejk} - 0.5)^2}{f_{ejk}} \tag{8.7}$$

Daniel (1990) has cited a number of criticisms to the Yates's continuity correction. While he recognizes that the procedure has been frequently used, he also observes a decline in its popularity. Toward the end of this chapter, we present an alternative for analyzing a 2 × 2 contingency table using the Fisher exact test.

At this point, the analysis is limited to identifying an association's presence or absence. In other words, the chi-square test's level of significance does not describe the strength of its association. We can use the *effect size* to analyze the degree of association. For the chi-square test for independence, the effect size between the nominal variables of a 2 × 2 contingency table can be calculated and represented with the phi (ϕ) coefficient (see Formula 8.8).

$$\phi = \sqrt{\frac{\chi^2}{n}} \tag{8.8}$$

where χ^2 is the chi-square test statistic and n is the number in the entire sample.

The phi coefficient ranges from 0 to 1. Cohen (1988) defined the conventions for effect size as small = 0.10, medium = 0.30, and large = 0.50. (Correlation coefficient and effect size are both measures of association. See Chapter 7 concerning correlation for more information on Cohen's assignment of effect size's relative strength.)

When the chi-square contingency table is larger than 2 × 2, Cramer's V statistic may be used to express effect size. The formula for Cramer's V is shown in Formula 8.9.

$$V = \sqrt{\frac{\chi^2}{(n)(L-1)}} \tag{8.9}$$

where χ^2 is the chi-square test statistic, n is the total number in the sample, and L is the minimum value of the row total and column total from the contingency table.

8.4.2 Sample Chi-Square Test for Independence

A counseling department for a school system is conducting a study to investigate the association between children's attendance in public and private preschools and their behavior in the kindergarten classroom. It is the researcher's desire to see if there is any positive association between early exposure to learning and behavior in the classroom.

The sample size for this study was $n = 100$. The data in Table 8.10 represent the observed frequencies for the 100 children whose behavior was observed during their

TABLE 8.10

	Behavior in Kindergarten			
	Poor	Average	Good	Row Total
Public preschool	12	25	10	**47**
Private preschool	6	12	0	**18**
No preschool	2	23	10	**35**
Column total	**20**	**60**	**20**	**100**

first 6 weeks of school. The students who were in the study were identified by the type of preschool educational exposure they received.

We want to determine if there is any association between type of preschool experience and behavior in kindergarten in the first 6 weeks of school. Since the data only need to be classified into categories, and no sample mean or sum of squares needs to be calculated, the chi-square statistic for independence can be used to test the nonparametric data.

1. *State the null and research hypotheses.*

    The null hypothesis, shown below, states that there is no association between the two categories. The behavior of the children in kindergarten is independent of the type of preschool experience they had. The research hypothesis states that there is a significant association between the preschool experience of the children and their behavior in kindergarten.

    The null hypothesis is

    H_0: In the general population, there is no association between type of preschool experience a child has and the child's behavior in kindergarten.

    The research hypothesis states

    H_A: In the general population there is a predictable relationship between the preschool experience and the child's behavior in kindergarten.

2. *Set the level of risk (or the level of significance) associated with the null hypothesis.*

    The level of risk, also called an alpha (α), is frequently set at 0.05. We will use an alpha of 0.05 in our example. In other words, there is a 95% chance that any observed statistical difference will be real and not due to chance.

3. *Choose the appropriate test statistic.*

    The data are obtained from 100 children in kindergarten who experienced differing preschool preparation prior to entering formal education. The

THE CHI-SQUARE TEST FOR INDEPENDENCE

kindergarten teachers for the children were asked to rate students' behaviors using three broad levels of ratings obtained from a survey. The students were then divided into three groups according to preschool experience (no preschool, private preschool, and public preschool). These data are organized into a two-dimensional categorical distribution that can be analyzed using an independent chi-square test.

4. *Compute the test statistic.*

First, tally the observed frequencies, f_{ojk}, for the 100 students who were in the study. Use these data to create the observed frequency table shown in Table 8.11.

TABLE 8.11

	Behavior in Kindergarten (Observed)			
	Poor	Average	Good	Row Total
Public preschool	12	25	10	**47**
Private preschool	6	12	0	**18**
No preschool	2	23	10	**35**
Column total	**20**	**60**	**20**	**100**

Next, calculate the expected frequency, f_{ejk}, for each category.

$$f_{e11} = \frac{(47)(20)}{100} = 940/100 = 9.4$$

$$f_{e12} = \frac{(47)(60)}{100} = 2820/100 = 28.2$$

$$f_{e13} = \frac{(47)(20)}{100} = 940/100 = 9.4$$

$$f_{e21} = \frac{(18)(20)}{100} = 360/100 = 3.6$$

$$f_{e22} = \frac{(18)(60)}{100} = 1080/100 = 10.8$$

$$f_{e23} = \frac{(18)(20)}{100} = 360/100 = 3.6$$

$$f_{e31} = \frac{(35)(20)}{100} = 700/100 = 7.0$$

$$f_{e32} = \frac{(35)(60)}{100} = 2100/100 = 21.0$$

$$f_{e33} = \frac{(35)(20)}{100} = 700/100 = 7.0$$

Place these values in an expected frequencies table (see Table 8.12).

Using the values for the observed and expected frequencies in Tables 8.11 and 8.12, the chi-square statistic can be calculated.

TABLE 8.12

	Behavior in Kindergarten (Expected)			
	Poor	Average	Good	Row Total
Public preschool	9.4	28.2	9.4	**47**
Private preschool	3.6	10.8	3.6	**18**
No preschool	7.0	21.0	7.0	**35**
Column total	**20**	**60**	**20**	**100**

$$\chi^2 = \sum_j \sum_k \frac{(f_{ojk} - f_{ejk})^2}{f_{ejk}}$$

$$= \frac{(12-9.4)^2}{9.4} + \frac{(25-28.2)^2}{28.2} + \frac{(10-9.4)^2}{9.4} + \frac{(6-3.6)^2}{3.6} + \frac{(12-10.8)^2}{10.8}$$

$$+ \frac{(0-3.6)^2}{3.6} + \frac{(2-7)^2}{7} + \frac{(23-21)^2}{21} + \frac{(10-7)^2}{7}$$

$$= \frac{(2.6)^2}{9.4} + \frac{(-3.2)^2}{28.2} + \frac{(0.6)^2}{9.4} + \frac{(2.4)^2}{3.6} + \frac{(1.2)^2}{10.8}$$

$$+ \frac{(-3.6)}{3.6} + \frac{(-5)^2}{7} + \frac{(2)^2}{21} + \frac{(3)^2}{7}$$

$$= 0.72 + 0.36 + 0.04 + 1.60 + 0.13 + 3.60 + 3.57 + 0.19 + 1.29$$

$$= 11.50$$

5. *Determine the value needed for rejection of the null hypothesis using the appropriate table of critical values for the particular statistic.*

    Before we go to the table of critical values, we need to determine the degrees of freedom, df. In this example, there are three categories in the preschool experience dimension, $R = 3$, and three categories in the behavior dimension, $C = 3$. To find the degrees of freedom, use df $= (R - 1)(C - 1) = (3 - 1)(3 - 1)$. Therefore, df $= 4$.

    Now, we use Table B.2, which lists the critical values for the chi-square. The critical value is found in the chi-square table for four degrees of freedom, df $= 4$. Since we set $\alpha = 0.05$, the critical value is 9.49. A calculated value that is greater than or equal to 9.49 will lead us to reject the null hypothesis.

6. *Compare the obtained value to the critical value.*

    The critical value for rejecting the null hypothesis is 9.49 and the obtained value is $\chi^2 = 11.50$. If the critical value is less than or equal to the obtained value, we must reject the null hypothesis. If instead, the critical value exceeds the obtained value, we do not reject the null hypothesis. Since the critical value is less than our obtained value, we must reject the null hypothesis.

7. *Interpret the results.*

We rejected the null hypothesis, suggesting that there is a real association between type of preschool experience children obtained and their behavior in the kindergarten classroom during their first few weeks in school. In particular, data tend to show that children who have private schooling do not tend to get good behavior ratings in school. The other area that tends to show some significant association is between poor behavior and no preschool experience. The students who had no preschool had very few poor behavior ratings in comparison to the other two groups.

At this point, the analysis is limited to identifying an association's presence or absence. In other words, the chi-square test's level of significance does not describe the strength of its association. The American Psychological Association (2001), however, has called for a measure of the degree of association called the *effect size*. For the chi-square test for independence with a 3 × 3 contingency table, we determine the strength of association, or the effect size, using Cramer's V.

From Table 8.11, we find that $L = 3$. For $n = 100$ and $\chi^2 = 11.50$, we use Formula 8.9 to determine V.

$$V = \sqrt{\frac{11.50}{(100)(3-1)}}$$
$$= \sqrt{0.06}$$
$$= 0.24$$

Our effect size, Cramer's V, is 0.24. This value indicates a medium level of association between type of preschool experience children obtained and their behavior in the kindergarten classroom during their first few weeks in school.

8. *Reporting the results.*

The reporting of results for the chi-square test for independence should include such information as the total number of participants in the sample and the number of participants classified in each of the categories. In addition, include the χ^2 statistic, degrees of freedom, and the *p*-value's relation to α. For this study, the number of children who were in each category (including preschool experience and behavior rating) should be presented in the two-dimensional table (see Table 8.10).

For this example, the records of 100 kindergarten students were examined to determine whether there was an association between preschool experience and behavior in kindergarten. The three preschool experiences were no preschool, private preschool, and public preschool. The three behavior ratings were poor, average, and good. The chi-square test was significant ($\chi^2_{(4)} = 11.50, p < 0.05$). Moreover, our effect size, using Cramer's V, was 0.24. Based on the results, there was a tendency shown for students with private preschool to not have good behavior and those with no preschool to not have poor behavior. It also indicated that average behavior was strong for all three preschool experiences.

8.4.3 Performing the Chi-Square Test for Independence Using SPSS

We will analyze the data from the above example using SPSS.

1. *Define your variables.*

   First, click the "Variable View" tab at the bottom of your screen, as shown in Figure 8.8. The chi-square test for independence requires variables to identify

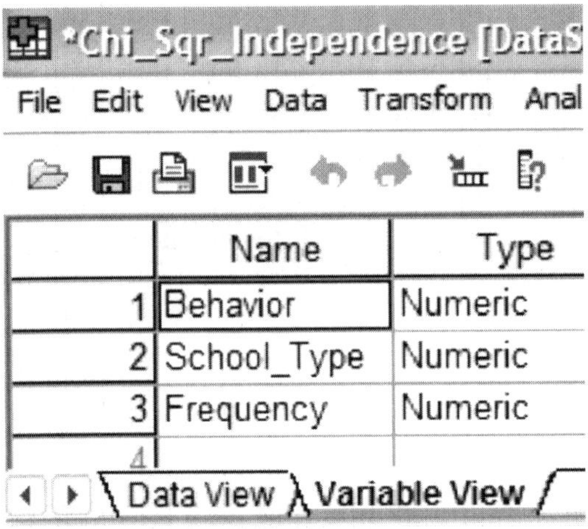

   FIGURE 8.8

   the conditions in the rows: one variable to identify the conditions of the rows and a second variable to identify the conditions of the columns. According to the previous example, the "Behavior" variable will represent the columns. "School_Type" will represent the rows. Finally, we need a variable to represent the observed frequencies. "Frequency" represents the observed frequencies.

   You must assign values to serve as a reference for the column and row variables. It is often easiest to assign each category a whole number value. First, click the gray square in the "Values" field to set the desired values. As shown in Figure 8.9, we have already assigned the value labels for the "Behavior" variable. For the "School_Type" variable, we set a value of 1 to equal "Public Preschool" and a value of 2 to equal "Private Preschool". As soon as we click the "Add" button, we will have set "No Preschool" equal to 3.

2. *Type in your values.*

   Click the "Data View" tab at the bottom of your screen, as shown in Figure 8.10. Use the whole number references you set earlier for the row and column variables. Each possible combination of conditions should exist. Then, enter the corresponding observed frequencies. In our example, row 1 represents a "Behavior" of 1, which is "Poor", and a "School_Type" of 1, which is "Public School". The observed frequency for this condition is 12.

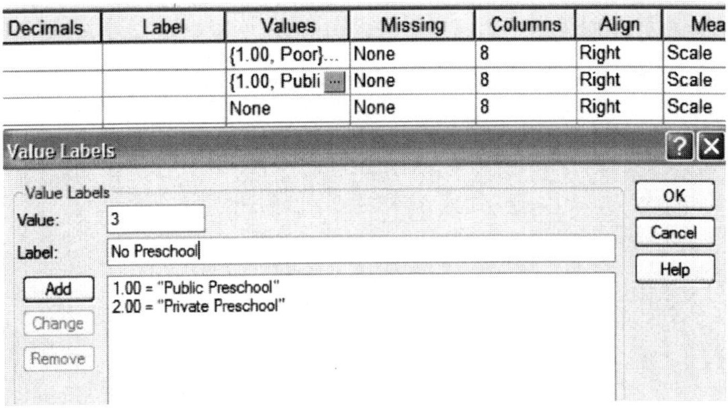

THE CHI-SQUARE TEST FOR INDEPENDENCE

FIGURE 8.9

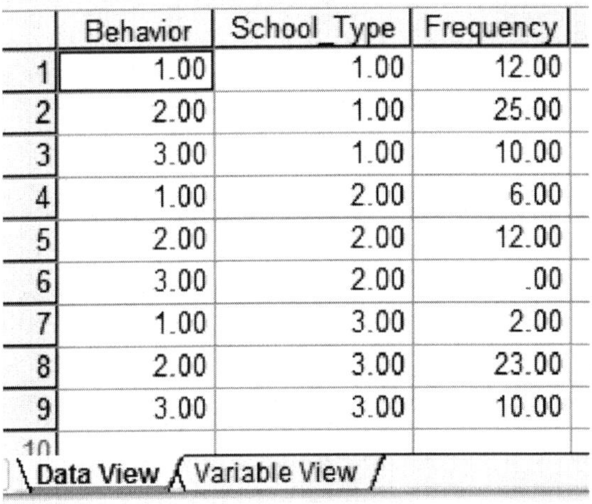

FIGURE 8.10

3. *Analyze your data.*

First, use the "Weight Cases" command to allow the observed frequencies variable to reference the categories variable. As shown in Figure 8.11, use the pull-down menus to choose "Data" and "Weight Cases...".

The default setting is "Do not weight cases". Click the circle next to "Weight cases by" as shown in Figure 8.12. Select the variable with the observed frequencies. Move that variable to the "Frequency Variable:" box by clicking the small arrow button. In our example, we have moved the "Frequency" variable. Finally, click "OK".

As shown in Figure 8.13, use the pull-down menus to choose "Analyze", "Descriptive Statistics", and "Crosstabs...".

TESTS FOR NOMINAL SCALE DATA: CHI-SQUARE AND FISHER EXACT TEST

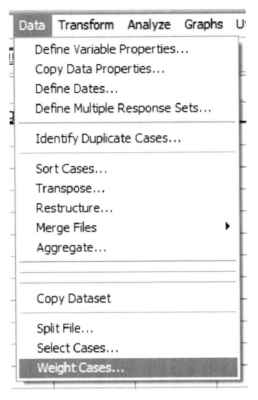

FIGURE 8.11

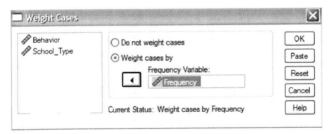

FIGURE 8.12

When the Crosstabs window is open, move the variable that represents the rows to the "Row(s):" box by selecting that variable and clicking the small arrow button next to that box. As shown in Figure 8.14, we have chosen the "School_Type" variable. Then, move the variable that represents the column to the "Column(s):" box. In our example, we have chosen the "Behavior" variable.

THE CHI-SQUARE TEST FOR INDEPENDENCE

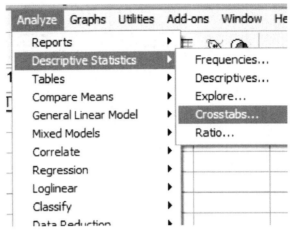

FIGURE 8.13

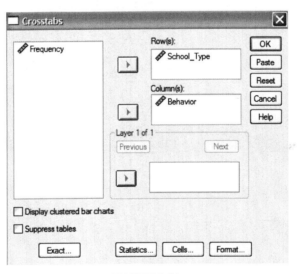

FIGURE 8.14

Next, click the "Statistics..." button. As shown in Figure 8.15, check the box next to "Chi-square" and the box next to "Phi and Cramer's V". Then, click "Continue" to return to the Crosstabs window.

Next, click the "Cells..." button. As shown in Figure 8.16, check the boxes next to "Observed" and "Expected". Then, click "Continue" to return to the Crosstabs window. Finally, click "OK" to perform the analysis.

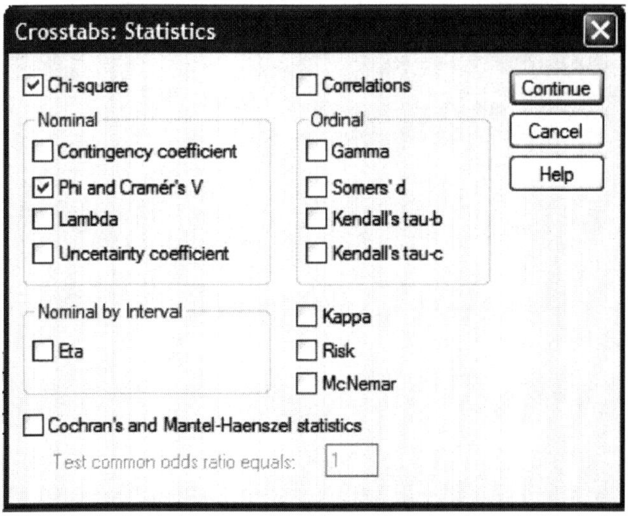

FIGURE 8.15

FIGURE 8.16

4. *Interpret the results from the SPSS Output window.*

School_Type * Behavior Crosstabulation

			Behavior			Total
			Poor	Average	Good	
School_Type	Public Preschool	Count	12	25	10	47
		Expected Count	9.4	28.2	9.4	47.0
	Private Preschool	Count	6	12	0	18
		Expected Count	3.6	10.8	3.6	18.0
	No Preschool	Count	2	23	10	35
		Expected Count	7.0	21.0	7.0	35.0
Total		Count	20	60	20	100
		Expected Count	20.0	60.0	20.0	100.0

The second, third, and fourth output tables from SPSS are of interest in this procedure. The second SPSS output table provides the observed and expected frequencies for each category and the total counts.

Chi-Square Tests

	Value	df	Asymp. Sig. (2-sided)
Pearson Chi-Square	11.502[a]	4	.021
Likelihood Ratio	16.042	4	.003
Linear-by-Linear Association	3.072	1	.080
N of Valid Cases	100		

a. 2 cells (22.2%) have expected count less than 5. The minimum expected count is 3.60.

The third SPSS output table provides the chi-square statistic ($\chi^2 = 11.502$), the degrees of freedom (df = 4), and the significance ($p = 0.021$).

Symmetric Measures

		Value	Approx. Sig.
Nominal by Nominal	Phi	.339	.021
	Cramer's V	.240	.021
N of Valid Cases		100	

The fourth SPSS output table provides the Cramer's V statistic ($V = 0.240$) to determine the level of association, or effect size.

Based on the results from SPSS, three programs were compared with unequal expected frequencies. The chi-square goodness-of-fit test was significant ($\chi^2_{(4)} = 11.502$, $p < 0.05$). Based on these results, there is a real association between type of preschool experience children obtained and their behavior in the kindergarten classroom during their first few weeks in school. In addition, the measured effect size presented a medium level of association ($V = 0.240$).

8.5 THE FISHER EXACT TEST

A special case arises if a contingency table's size is 2×2 and at least one expected cell count is less than 5. In this circumstance, SPSS will calculate a Fisher exact test instead of a chi-square test for independence.

The Fisher exact test is useful for analyzing discrete data obtained from small, independent samples. They can be either nominal or ordinal. It is used when the scores of two independent random samples fall into one of the two mutually exclusive classes or one of the two possible scores is obtained. The results form a 2×2 contingency table, as noted earlier.

In this chapter, we will describe how to perform and interpret the Fisher exact test for different samples.

8.5.1 Computing the Fisher Exact Test for 2 × 2 Tables

Compare the 2 × 2 contingency table's one-sided significance with the level of risk, α. Table 8.13 is the 2 × 2 contingency table that is used as the basis for computing Fisher exact test's one-sided significance.

TABLE 8.13

Variable	Group I	Group II	Combined
+	A	B	A + B
−	C	D	C + D
Total	A + C	B + D	N

The formula for computing the one-sided significance for the Fisher exact test is shown in Formula 8.10. Table B.9 lists the factorials for $n = 0$ to $n = 25$.

$$p = \frac{(A+B)!(C+D)!(A+C)!(B+D)!}{N!A!B!C!D!} \tag{8.10}$$

If all cell counts are equal to or larger than 5 ($n_i \geq 5$), Daniel (1990) suggested that one use a large sample approximation with the chi-square test instead of the Fisher exact test.

8.5.2 Sample Fisher Exact Test

A small medical center administered a survey to determine its nurses' attitude of readiness to care for patients. The survey was a 15-item Likert scale with two points positive, two points negative, and a neutral point. The study was conducted to compare the feelings between men and women. Each person was classified according to a total attitude determined by summing the item values on the survey. A maximum positive attitude would be $+33$ and a maximum negative attitude would be -33.

Tables 8.14 and 8.15 show the number of men and women who had positive and negative attitudes about how they were prepared. There were five men and six women. Four of the men had positive survey results and only one of the women had a positive survey result.

We want to determine if there is a difference in attitude between men and women toward their preparation to care for patients. Since the data form a 2 × 2 contingency table and at least one cell has an expected count (see Formula 8.2) of less than 5, the Fisher exact test is a useful procedure to analyze the data and test the hypothesis.

1. *State the null and research hypotheses.*

    The null hypothesis, shown below, states that there are no differences between men and women on the attitude survey that measures feelings about

TABLE 8.14

Participant	Gender	Score	Attitude
1	Male	+30	+
2	Male	+14	+
3	Male	−21	−
4	Male	+22	+
5	Male	+9	+
6	Female	−22	−
7	Female	−13	−
8	Female	−20	−
9	Female	−7	−
10	Female	+19	+
11	Female	−31	−

TABLE 8.15

	Group		
	Men	Women	
Positive	4	1	5
Negative	1	5	6
	5	6	11

the program that teaches care for patients. The alternative hypothesis is that the proportion of men with positive attitudes, P_M, exceeds the proportion of women with positive attitudes, P_W.

The hypotheses can be written as shown below.

$H_0: P_M = P_W$
$H_A: P_M > P_W$

2. *Set the level of risk (or the level of significance) associated with the null hypothesis.*

    The level of risk, also called an alpha (α) level, is frequently set at 0.05. We will use an alpha of 0.05 in our example. In other words, there is a 95% chance that any observed statistical difference will be real and not due to chance.

3. *Choose the appropriate test statistic.*

    The data are obtained from a 2 × 2 contingency table. Two independent groups were measured on a survey and classified according to two criteria. The classifications were (+) for a positive attitude and (−) for negative attitude. The samples are small, thus requiring nonparametric statistics. We are analyzing data in a 2 × 2 contingency table and at least one cell has an expected count (see Formula 8.2) of less than 5. Therefore, we will use the Fisher exact test.

4. *Compute the test statistic.*

    First, construct the 2 × 2 contingency tables for the data in the study and for data that represent a more extreme occurrence than that obtained. In this example, there were five men and six women who were in the training program for nurses. Four of the men responded positively to the survey and only one of the women responded positively. The remainder of the people in the study responded negatively.

If we wish to test the null hypothesis statistically, we must consider the possibility of the occurrence of the more extreme outcome that is shown in Table 8.16b. In that table, none of the men responded negatively, and none of the women responded positively.

The tables for the statistic are given below. Table 8.16a shows the results that occurred in the data collected and Table 8.16b shows the more extreme outcome that could occur.

TABLE 8.16a

	Group		
	Men	Women	
Positive	4	1	5
Negative	1	5	6
	5	6	11

TABLE 8.16b

	Group		
	Men	Women	
Positive	5	0	5
Negative	0	6	6
	5	6	11

To test the hypothesis, we first use Formula 8.8 to compute the probability of each possible outcome shown above.

For Table 8.16a,

$$p_1 = \frac{(A+B)!(C+D)!(A+C)!(B+D)!}{N!A!B!C!D!}$$

$$= \frac{(4+1)!(1+5)!(4+1)!(1+5)!}{(11!)(4!)(1!)(1!)(5!)}$$

$$= \frac{(5!)(6!)(5!)(6!)}{(11!)(4!)(1!)(1!)(5!)}$$

$$= \frac{(120)(720)(120)(720)}{(39916800)(24)(1)(1)(120)}$$

$$= 0.065$$

THE FISHER EXACT TEST

For Table 8.16b,

$$p_2 = \frac{(A+B)!(C+D)!(A+C)!(B+D)!}{N!A!B!C!D!}$$

$$= \frac{(5+0)!(0+6)!(5+0)!(0+6)!}{(11!)(5!)(0!)(0!)(6!)}$$

$$= \frac{(5!)(6!)(5!)(6!)}{(11!)(5!)(0!)(0!)(6!)}$$

$$= \frac{(120)(720)(120)(720)}{(39916800)(120)(1)(1)(720)}$$

$$= 0.002$$

The probability is found by adding the two results that were computed above.

$$p = p_1 + p_2 = 0.065 + 0.002$$
$$= 0.067$$

5. *Determine the value needed for rejection of the null hypothesis using the appropriate table of critical values for the particular statistic.*

   In the example in this chapter, the probability was computed and compared to the level of risk specified earlier, $\alpha = 0.05$. This computational process involves very large numbers and is aided by the table values. It is recommended that a table of critical values be used when possible.

6. *Compare the obtained value to the critical value.*

   The critical value for rejecting the null hypothesis is $\alpha = 0.05$ and the obtained p-value is $p = 0.067$. If the critical value is greater than the obtained value, we must reject the null hypothesis. If the critical value is less than the obtained value, we do not reject the null hypothesis. Since the critical value is less than the obtained value, we do not reject the null hypothesis.

7. *Interpret the results.*

   We did not reject the null hypothesis, suggesting that no real difference existed between the attitudes of men and women about their readiness to care for patients. There was, however, a strong trend toward positive feelings on the part of the men and negative feelings on the part of the women. The probability was small, although not significant. This is the type of study that would call for further investigation with other samples to see if this trend was more pronounced. Our analysis does provide some evidence that there is some difference, and if analyzed with a more liberal critical value, of say $\alpha = 0.10$, this statistical test would show significance.

   Since the Fisher exact test was not statistically significant ($p > \alpha$), we may not have an interest in the strength of the association between the two variables. However, a researcher wishing to replicate the study may wish to know that strength of association.

The *effect size* is a measure of association between two variables. For the Fisher exact test, which has a 2 × 2 contingency table, we determine the effect size using the phi (ϕ) coefficient (see Formula 8.8). For $n = 11$ and $\chi^2 = 4.412$ (calculation for χ^2 not shown), we use Formula 8.8 to determine ϕ.

$$\phi = \sqrt{\frac{4.412}{11}}$$

$$= \sqrt{0.401}$$

$$= 0.633$$

Our effect size, the phi (ϕ) coefficient, is 0.633. This value indicates a strong level of association between the two variables. What is more, a replication of the study may be worth the effort.

8. *Reporting the results.*

When reporting the findings include the table that shows the actual reported frequencies, including all marginal frequencies. In addition, report the *p*-value and its relationship to the critical value.

For this example, Table 8.15 would be reported. The obtained significance, $p = 0.067$, was greater than the critical value, $\alpha = 0.05$. Therefore, we did not reject the null hypothesis, suggesting that there was no difference between men and women on the attitude survey that measures feelings about the program that teaches care for patients.

8.5.3 Performing the Fisher Exact Test Using SPSS

As noted earlier, SPSS performs a Fisher exact test instead of a chi-square test for independence if the contingency table's size is 2 × 2 and at least one expected cell count is less than 5. In other words, to perform a Fisher exact test, use the same method you used for a chi-square test for independence.

The SPSS Outputs provide the results for the sample Fisher exact test computed earlier. Note that all four expected counts were less than 5. In addition, the one-sided significance is $p = 0.067$.

The last SPSS Output provides the effect size for the association. Since the association was not statistically significant ($p > \alpha$), the effect size ($\phi = 0.633$) was not of interest to this study.

Attitude * Gender Crosstabulation

			Gender		Total
			male	female	
Attitude	positive attitude	Count	4	1	5
		Expected Count	2.3	2.7	5.0
	negative attitude	Count	1	5	6
		Expected Count	2.7	3.3	6.0
Total		Count	5	6	11
		Expected Count	5.0	6.0	11.0

Chi-Square Tests

	Value	df	Asymp. Sig. (2-sided)	Exact Sig. (2-sided)	Exact Sig. (1-sided)
Pearson Chi-Square	4.412[b]	1	.036		
Continuity Correction[a]	2.228	1	.136		
Likelihood Ratio	4.747	1	.029		
Fisher's Exact Test				.080	.067
Linear-by-Linear Association	4.011	1	.045		
N of Valid Cases	11				

a. Computed only for a 2x2 table
b. 4 cells (100.0%) have expected count less than 5. The minimum expected count is 2.27.

Symmetric Measures

		Value	Approx. Sig.
Nominal by Nominal	Phi	.633	.036
	Cramer's V	.633	.036
N of Valid Cases		11	

8.6 EXAMPLES FROM THE LITERATURE

Below are varied examples of the nonparametric procedures described in this chapter. We have summarized each study's research problem and researchers' rationale(s) for choosing a nonparametric approach. We encourage you to obtain these studies if you are interested in their results.

- Duffy, R. D., & Sedlacek, W. E. (2007). The work values of first-year college students: Exploring group differences. *The Career Development Quarterly, 55* (4), 359–364.

  Duffy and Sedlacek examined the surveys of 3570 first-year college students regarding the factors they deemed most important to their long-term career choice. Chi-square analyses were used to assess work value differences (social, extrinsic, prestige, and intrinsic) in gender, parental income, race, and educational aspirations. The researchers used chi-square tests for independence since these data were frequencies of nominal items.

- Ferrari, J. R., Athey, R. B., Moriarty, M. O., & Appleby, D. C. (2006). Education and employment among alumni academic honor society leaders. *Education, 127* (2), 244–259.

  Ferrari, Athey, Moriarty, and Appleby studied the effects of leadership experience in an academic honor society on later employment and education. Drawing from honor society alumni, the researchers compared leaders with non-leaders on various aspects of their graduate education or employment. Since most data were frequencies of nominal items, the researchers used chi-square test for independence.

- Helsen, W., Gilis, B., & Weston, M. (2006). Errors in judging "offside" in association football: Test of the optical error versus the perceptual flash-lag hypothesis. *Journal of Sports Sciences, 24*(5), 521–528.

Helsen, Gilis, and Weston analyzed the correctness of assistant referees' offside judgments during the final round of the FIFA 2002 World Cup. Specifically, they used digital video technology to examine situations involving the viewing angle and special position of a moving object. They used a chi-squared goodness-of-fit test to determine if the ratio of correct to incorrect decisions and the total number of offside decisions were uniformly distributed throughout six 15 min. intervals. They also used a chi-squared goodness-of-fit test to determine if flag errors versus non-flag errors led to a judgment bias.

- Shorten, J., Seikel, M., & Ahrberg, J. H. (2005). Why do you still use Dewey? Academic libraries that continue with Dewey decimal classification. *Library Resources & Technical Services, 49*(2), 123–136.

    Shorten, Seikel, and Ahrberg analyzed a survey of 34 academic libraries in the United States and Canada that use the Dewey Decimal Classification (DDC). They wished to determine why these libraries continue using DDC and if they have considered reclassification. Some of the survey questions asked participants to respond to a reason by selecting "more important", "less important", or "not a reason at all". Responses were analyzed with a chi-squared goodness-of-fit test to examine responses for consensus among libraries.

- Rimm-Kaufman, S. E., & Zhang, Y. (2005). Father–school communication in preschool and kindergarten, *The School Psychology Review, 34*(3), 287–308.

    Rimm-Kaufman and Zhang studied the communication between fathers of "at-risk" children and their preschool and kindergarten schools. Specifically, they examined frequency, characteristics, and predictors of communication based on family sociodemographic characteristics. When analyzing frequencies, they used chi-squared tests. However, when cells contained frequencies of zero, they used a Fisher exact test.

- Enander, R. T., Gagnon, R. N., Hanumara, R. C., Park, E., Armstrong, T., & Gute, D. M. (2007). Environmental health practice: Statistically based performance measurement. *American Journal of Public Health, 97*(5), 819–824.

    Enander, Gagnon, Hanumara, Park, Armstrong, and Gute investigated a newly employed inspection method for self-certification of environmental and health qualities in automotive refinishing facilities. They focused on occupational health and safety, air pollution control, hazardous waste management, and wastewater discharge. A Fisher exact test was used to analyze 2×2 tables with relatively small observed cell frequencies.

- Mansell, S., Sobsey, D., & Moskal, R. (1998). Clinical findings among sexually abused children with and without developmental disabilities. *Mental Retardation, 36*(1), 12–22.

    To examine the clinical problems of sexual abuse, Mansell, Sobsey, and Moskal compared children with developmental disabilities to children without developmental disabilities. Categorical data were analyzed with chi-square tests of independence; however, the Yates' continuity corrections were used when cell counts exhibited less than the minimum expected count needed for the chi-square test.

8.7 SUMMARY

Nominal, or categorical, data sometimes need analyses. In such cases, you may be seeking to determine if the data statistically match some known or expected set of frequencies. Or, you may wish to determine if two or more categories are statistically independent. In either case, nominal data can be analyzed with a nonparametric procedure.

In this chapter, we presented three procedures for examining nominal data: chi-square (χ^2) goodness-of-fit, chi-square test for independence, and the Fisher exact test. We also explained how to perform the procedures using SPSS. Finally, we offered varied examples of these nonparametric statistics from the literature. In the next chapter, we will describe how to determine if a series of events occurred randomly.

8.8 PRACTICE QUESTIONS

1. A police department wishes to compare the average number of monthly robberies at four locations in their town. Use equal categories to identify one or more concentrations of robberies. The data are presented in Table 8.17.

   TABLE 8.17

	Average Monthly Robberies
Location 1	15
Location 2	10
Location 3	19
Location 4	16

   Use a chi-square goodness-of-fit test with $\alpha = 0.05$ to determine if the robberies are concentrated in one or more of the locations. Report your findings.

2. The chi-square goodness-of-fit test serves as a useful tool to ensure that statistical samples approximately match the desired stratification proportions of the population from which they are drawn.

   A researcher wishes to determine if her randomly drawn sample matches the racial stratification of school age children. She used the most recent U.S. Census data, which was from 2001. The racial composition of her sample and the 2001 U.S. Census proportions are displayed in Table 8.18.

   Use a chi-square goodness-of-fit test with $\alpha = 0.05$ to determine if the researcher's sample matches the proportions reported by the U.S. Census. Report your findings.

3. A researcher wishes to determine if there is an association between the level of a teacher's education and his/her job satisfaction. He surveyed 158 teachers. The frequencies of the corresponding results are displayed in Table 8.19.

TABLE 8.18

Race	Frequency of Race from the Researcher's Randomly Drawn Sample	Racial Percentage of U.S. School Children Based on the 2001 U.S. Census
White	57	72%
Black	21	20%
Asian, Hispanic, or Pacific Islander	14	8%

TABLE 8.19

	Teacher Education Level (Observed)			
	Bachelor Degree	Master Degree	Post-Master Degree	Row Total
Satisfied	60	41	19	**120**
Unsatisfied	10	13	15	**38**
Column total	**70**	**54**	**34**	**158**

First, use a chi-square test for independence with $\alpha = 0.05$ to determine if there is an association between level of education and job satisfaction. Then, determine the effect size for the association. Report your findings.

4. A professor gave her class a 10-item survey to determine the students' satisfaction with the course. Survey question responses were measured using a five-point Likert scale. The survey had a score range of $+20$ to -20. Table 8.20 shows the scores of the students in a class of 13 students who rated the professor.

TABLE 8.20

Participant	Gender	Score	Satisfaction
1	Male	+12	+
2	Male	+6	+
3	Male	−5	−
4	Male	−10	−
5	Male	+17	+
6	Male	+4	+
7	Female	−2	−
8	Female	−13	−
9	Female	+10	+
10	Female	−8	−
11	Female	−11	−
12	Female	−4	−
13	Female	−14	−

SOLUTIONS TO PRACTICE QUESTIONS 189

Use a Fisher exact test with $\alpha = 0.05$ to determine if there is an association between gender and course satisfaction of the professor's class. Then, determine the effect size for the association. Report your findings.

8.9 SOLUTIONS TO PRACTICE QUESTIONS

1. The results from the analysis are displayed in the SPSS Outputs below.

Location

	Observed N	Expected N	Residual
Location 1	15	15.0	.0
Location 2	10	15.0	-5.0
Location 3	19	15.0	4.0
Location 4	16	15.0	1.0
Total	60		

Test Statistics

	Location
Chi-Square[a]	2.800
df	3
Asymp. Sig.	.423

a. 0 cells (.0%) have expected frequencies less than 5. The minimum expected cell frequency is 15.0.

According to the data, the results from the chi-square goodness-of-fit test were not significant ($\chi^2_{(3)} = 2.800$, $p > 0.05$). Therefore, no particular location displayed a significantly higher or lower number of robberies.

2. The results from the analysis are displayed in the SPSS Outputs below.

Race

	Observed N	Expected N	Residual
White	57	66.2	-9.2
Black	21	18.4	2.6
Asian, Hispanic, or Pacific Islander	14	7.4	6.6
Total	92		

Test Statistics

	Race
Chi-Square[a]	7.647
df	2
Asymp. Sig.	.022

a. 0 cells (.0%) have expected frequencies less than 5. The minimum expected cell frequency is 7.4.

According to the data, the results from the chi-square goodness-of-fit test were significant ($\chi^2_{(2)} = 7.647$, $p < 0.05$). Therefore, the sample's racial

stratification approximately matches the U.S. Census racial composition of school aged children in 2001.
3. The results from the analysis are displayed in the SPSS Outputs below.

Job_Satisfaction * Education_Level Crosstabulation

			Education_Level			Total
			Bachelor Degree	Master Degree	Post-Master Degree	
Job_Satisfaction	Satisfied	Count	60	41	19	120
		Expected Count	53.2	41.0	25.8	120.0
	Unsatisfied	Count	10	13	15	38
		Expected Count	16.8	13.0	8.2	38.0
Total		Count	70	54	34	158
		Expected Count	70.0	54.0	34.0	158.0

Chi-Square Tests

	Value	df	Asymp. Sig. (2-sided)
Pearson Chi-Square	11.150a	2	.004
Likelihood Ratio	10.638	2	.005
Linear-by-Linear Association	10.593	1	.001
N of Valid Cases	158		

a. 0 cells (.0%) have expected count less than 5. The minimum expected count is 8.18.

Symmetric Measures

		Value	Approx. Sig.
Nominal by Nominal	Phi	.266	.004
	Cramer's V	.266	.004
N of Valid Cases		158	

As seen in the first SPSS Output, none of the cells had an expected count of less than 5. Therefore, the chi-square test was indeed an appropriate analysis. Concerning effect size, the size of the contingency table was larger than 2 × 2. Therefore, a Cramer's V was appropriate.

According to the data, the results from the chi-square test for independence were significant ($\chi^2_{(2)} = 11.150$, $p < 0.05$). Therefore, the analysis provides evidence that teacher education level differentiates between individuals based on job satisfaction. In addition, the effect size ($V = 0.266$) indicated a medium level of association between the variables.

4. The results from the analysis are displayed in the SPSS Outputs below.

Satisfaction * Gender Crosstabulation

			Gender		Total
			male	female	
Satisfaction	positive	Count	4	1	5
		Expected Count	2.3	2.7	5.0
	negative	Count	2	6	8
		Expected Count	3.7	4.3	8.0
Total		Count	6	7	13
		Expected Count	6.0	7.0	13.0

Chi-Square Tests

	Value	df	Asymp. Sig. (2-sided)	Exact Sig. (2-sided)	Exact Sig. (1-sided)
Pearson Chi-Square	3.745[b]	1	.053		
Continuity Correction[a]	1.859	1	.173		
Likelihood Ratio	3.943	1	.047		
Fisher's Exact Test				.103	.086
Linear-by-Linear Association	3.457	1	.063		
N of Valid Cases	13				

a. Computed only for a 2x2 table
b. 4 cells (100.0%) have expected count less than 5. The minimum expected count is 2.31.

Symmetric Measures

		Value	Approx. Sig.
Nominal by Nominal	Phi	.537	.053
	Cramer's V	.537	.053
N of Valid Cases		13	

As seen in the first SPSS Output, all of the cells had an expected count of less than 5. Therefore, the Fisher exact test was an appropriate analysis. Concerning effect size, the size of the contingency table was 2×2. Therefore, a phi (ϕ) coefficient was appropriate.

According to the data, the results from the Fisher exact test were not significant ($p = 0.086$) based on $\alpha = 0.05$. Therefore, the analysis provides evidence that no association exists between gender and course satisfaction of the professor's class. In addition, the effect size ($\phi = 0.537$) was not of interest to this study due to the lack of significant association between variables.

9

TEST FOR RANDOMNESS: THE RUNS TEST

9.1 OBJECTIVES

In this chapter, you will learn the following items.

- How to use a runs test to analyze a series of events for randomness.
- How to perform a runs test using SPSS.

9.2 INTRODUCTION

Every investor wishes he or she could predict the behavior of a stock's performance. Is there a pattern to a stock's gain/loss cycle or are the events random? One could make a defensible argument to that question with an analysis of randomness.

The runs test (sometimes called a Wald–Wolfowitz runs test) is a statistical procedure for examining a series of events for randomness. This nonparametric test has no parametric equivalent. In this chapter, we will describe how to perform and interpret a runs test for both small samples and large samples. We will also explain how to perform the procedure using SPSS. Finally, we offer varied examples of these nonparametric statistics from the literature.

Nonparametric Statistics for Non-Statisticians, Gregory W. Corder and Dale I. Foreman
Copyright © 2009 John Wiley & Sons, Inc.

9.3 THE RUNS TEST FOR RANDOMNESS

The runs test seeks to determine if a series of events occur randomly or are merely due to chance. To understand a run, consider a sequence represented by two symbols, A and B. One simple example might be several tosses of a coin where A = heads and B = tails. Another example might be whether an animal chooses to eat first or drink first. Use A = eat and B = drink.

The first steps are to list the events in sequential order and count the number of runs. A *run* is a sequence of the same event written one or more times. For example, compare two event sequences. The first sequence is written AAAAAABBBBBB.

Then, separate the sequence into same groups as shown in Figure 9.1. There are two runs in this example, $R = 2$. This is a trend pattern in which events are clustered and it does not represent random behavior.

$$\underline{\text{AAAAAA}} \quad \underline{\text{BBBBBB}}$$
$$1 2$$

FIGURE 9.1

Consider a second event sequence written ABABABABABAB. Again, separate the events into same groups (see Figure 9.2) to determine the number of runs. There are 12 runs in this example, $R = 12$. This is a cyclic pattern and does not represent random behavior either. As illustrated in the two examples above, too few or too many runs lack randomness.

$$\underline{\text{A}}\ \underline{\text{B}}\ \underline{\text{A}}\ \underline{\text{B}}\ \underline{\text{A}}\ \underline{\text{B}}\ \underline{\text{A}}\ \underline{\text{B}}\ \underline{\text{A}}\ \underline{\text{B}}\ \underline{\text{A}}\ \underline{\text{B}}$$
$$1\ \ 2\ \ 3\ \ 4\ \ 5\ \ 6\ \ 7\ \ 8\ \ 9\ \ 10\ \ 11\ \ 12$$

FIGURE 9.2

A run can also describe how a sequence of events occurs in relation to a custom value. Use two symbols, such as A and B, to define whether an event exceeds or falls below the custom value. A simple example may reference the freezing point of water where A = temperatures above 0°C and B = temperatures below 0°C. In this example, simply list the events in order and determine the number of runs as described above.

After the number of runs is determined, it must be examined for significance. We may use a table of critical values (see Table B.10). However, if the number of values in each sample, n_1 or n_2, exceeds those available from the table, then a large sample approximation may be performed. For large samples, compute a z-score and use a table with the normal distribution (see Table B.1) to obtain a critical region of z-scores. Formulas 9.1–9.5 are used to find the z-score of a runs test for large samples.

$$\bar{x}_R = \frac{2n_1 n_2}{n_1 + n_2} + 1 \tag{9.1}$$

where $\bar{x}_R$ is the mean value of runs, n_1 is the number of times the first event occurred, and n_2 is the number of times the second event occurred.

$$s_R = \sqrt{\frac{2n_1 n_2 (2n_1 n_2 - n_1 - n_2)}{(n_1 + n_2)^2 (n_1 + n_2 - 1)}} \quad (9.2)$$

where s_R is the standard deviation of runs.

$$z^* = \frac{R + h - \bar{x}_R}{s_R} \quad (9.3)$$

where z^* is the z-score for a normal approximation of the data, R is the number of runs, and h is the correction for continuity, ± 0.5, where

$$h = +0.5 \quad \text{if } R < (2n_1 n_2/(n_1 + n_2) + 1) \quad (9.4)$$

and

$$h = -0.5 \quad \text{if } R > (2n_1 n_2/(n_1 + n_2) + 1) \quad (9.5)$$

9.3.1 Sample Runs Test (Small Data Samples)

The following study seeks to examine gender bias in science instruction. A male science teacher was observed during a typical class discussion. The observer noted the gender of the student that the teacher called on to answer a question. In the course of 15 minutes, the teacher called on 10 males and 10 females. The observer noticed that the science teacher called on equal numbers of males and females, but he wanted to examine the data for a pattern. To determine if the teacher used a random order to call on students with regard to gender, he used a runs test for randomness. Using M for male and F for female, the sequence of student recognition by the teacher is MFFMFMFMFFMFFFMMFMMM.

1. *State the null and research hypotheses.*

    The null hypothesis, shown below, states that the sequence of events is random. The research hypothesis, shown below, states that the sequence of events is not random.

    The null hypothesis is

    H_0: The sequence in which the teacher calls on males and females is random.

    The research hypothesis is

    H_A: The sequence in which the teacher calls on males and females is not random.

2. *Set the level of risk (or the level of significance) associated with the null hypothesis.*

    The level of risk, also called an alpha (α), is frequently set at 0.05. We will use an alpha of 0.05 in our example. In other words, there is a 95% chance that any observed statistical difference will be real and not due to chance.

3. *Choose the appropriate test statistic.*

    The observer is examining the data for randomness. Therefore, he is using a runs test for randomness.

THE RUNS TEST FOR RANDOMNESS

4. *Compute the test statistic.*

   First, determine the number of runs, R. It is helpful to separate the events as shown in Figure 9.3. The number of runs in the sequence is $R = 13$.

M	FF	M	F	M	F	M	FF	M	FFF	MM	F	MMM
1	2	3	4	5	6	7	8	9	10	11	12	13

   FIGURE 9.3

5. *Determine the value needed for rejection of the null hypothesis using the appropriate table of critical values for the particular statistic.*

   Since the sample sizes are small, we refer to Table B.10, which lists the critical values for the runs test. There were 10 males (n_1) and 10 females (n_2). The critical values are found in the table at the point for $n_1 = 10$ and $n_2 = 10$. We set $\alpha = 0.05$. The critical region for the runs test is $6 < R < 16$. If the number of runs, R, is 6 or less or 16 or greater, we reject our null hypothesis.

6. *Compare the obtained value to the critical value.*

   We found that $R = 13$. This value is within our critical region ($6 < R < 16$). Therefore, we do not reject the null hypothesis.

7. *Interpret the results.*

   We did not reject the null hypothesis, suggesting that the sequence of events is random. Therefore, we can state that the order in which the science teacher calls on males and females is random.

8. *Reporting the results.*

   The reporting of results for the runs test should include such information as the sample sizes for each group, the number of runs, and the p-value with respect to α.

   For this example, the runs test indicated that the sequence was random ($R = 13, n_1 = 10, n_2 = 10, p > 0.05$). Therefore, the study provides evidence that the science teacher was demonstrating no gender bias.

9.3.2 Performing the Runs Test Using SPSS

We will analyze the data from the above example using SPSS.

1. *Define your variables.*

   First, click the "Variable View" tab at the bottom of your screen. Then, type the names of your variables in the "Name" column. As seen in Figure 9.4, we call our variable "gender".

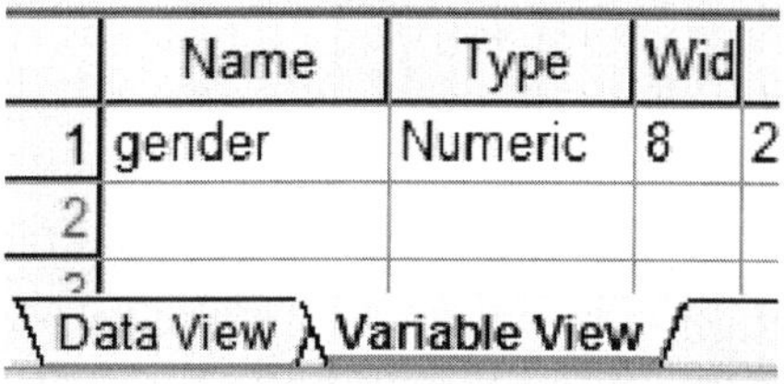

	Name	Type	Wid	
1	gender	Numeric	8	2
2				

   Data View \ Variable View

   FIGURE 9.4

Next, we establish a grouping variable to differentiate between males and females. When establishing a grouping variable, it is often easiest to assign each group a whole number value. As shown in Figure 9.5, our groups are "male" and "female". First, we select the "Values" column and click the gray square. Then, we set a value of 0 to equal "male". Now, as soon as we click the "Add" button, we will set "female" equal to 2 based on the values we inserted above. We did not choose the value of 1 since we will use it in step 3 as a reference (custom value) to compare the events.

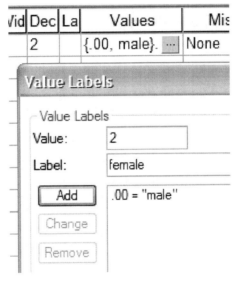

FIGURE 9.5

2. *Type in your values.*

   Click the "Data View" tab at the bottom of your screen (see Figure 9.6). Type the values into the column in the same order they occurred. Remember that we type 0 for "male" and 2 for "female".

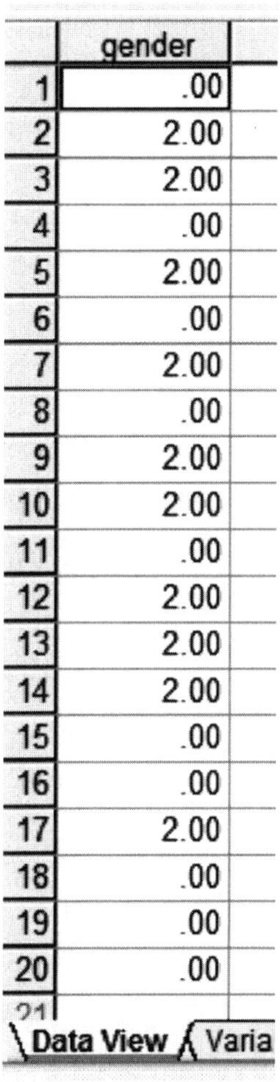

THE RUNS TEST FOR RANDOMNESS

	gender
1	.00
2	2.00
3	2.00
4	.00
5	2.00
6	.00
7	2.00
8	.00
9	2.00
10	2.00
11	.00
12	2.00
13	2.00
14	2.00
15	.00
16	.00
17	2.00
18	.00
19	.00
20	.00

FIGURE 9.6

3. *Analyze your data.*

As shown in Figure 9.7, use the pull-down menus to choose "Analyze", "Nonparametric Tests", and "Runs...".

The runs test required a reference point to compare the events. As shown in Figure 9.8 under "Cut Point", uncheck "Median" and check the box next to "Custom:". Type a value in the box that is between the events' assigned values. For our example, we used 0 and 2 for the events' values, so type a custom value of 1. Use the arrow button to place the variable with your data values in the box labeled "Test Variable List". In our example, we chose the variable "gender". Finally, click "OK" to perform the analysis.

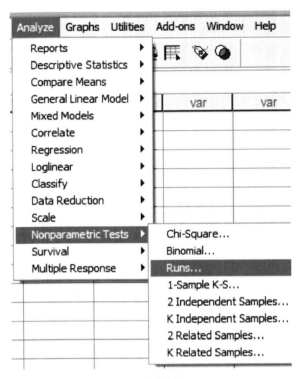

FIGURE 9.7

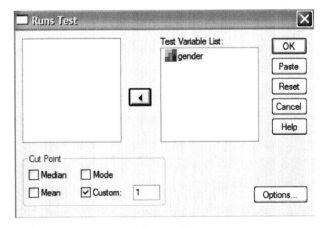

FIGURE 9.8

4. *Interpret the results from the SPSS Output window.*

Runs Test

	gender
Test Value[a]	1.0000
Total Cases	20
Number of Runs	13
Z	.689
Asymp. Sig. (2-tailed)	.491

a. User-specified.

The runs test SPSS output table returns the total number of observations ($N = 20$) and the number of runs ($R = 13$). SPSS also calculates the z-score ($z^* = 0.689$) and the two-tailed significance ($p = 0.491$).

5. *Determine the observation frequencies for each event.*

In order to determine the number of observations for each event, an additional set of steps is required. As shown in Figure 9.9, use the pull-down menus to choose "Analyze", "Descriptive Statistics", and "Frequencies... ".

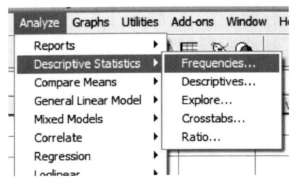

FIGURE 9.9

Next, use the arrow button to place the variable with your data values in the box labeled "Variable(s):" as shown in Figure 9.10. In our example, we chose the variable "gender". Finally, click "OK" to perform the analysis.

gender

		Frequency	Percent	Valid Percent	Cumulative Percent
Valid	male	10	50.0	50.0	50.0
	female	10	50.0	50.0	100.0
	Total	20	100.0	100.0	

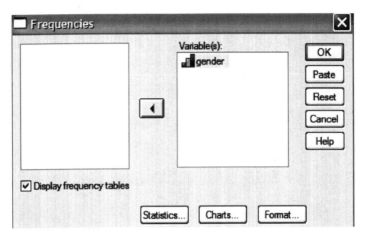

FIGURE 9.10

The second SPSS output table displays the frequencies for each event. Based on the results from SPSS, the runs test indicated that the sequence was random ($R = 13$, $n_1 = 10$, $n_2 = 10$, $p > 0.05$). Therefore, the science teacher randomly chose between males and females when calling on students.

9.3.3 Sample Runs Test (Large Data Samples)

The previous study investigating gender bias was replicated. This time, however, a different male teacher was observed and the observation occurred over a longer period of time. As before, the observer noted the gender of the student that the teacher called on to answer a question. In the course of 30 minutes, the teacher called on 23 males and 14 females. We will once again examine the data for a pattern and use a runs test to examine student recognition with respect to gender. This time, however, we will use a large sample approximation since at least one sample size is large. Using M for male and F for female, the sequence of student recognition by the teacher is FFMMFFFMFFFMMFMMMMFMMMMMMFMFMMFMMMMMF.

1. *State the null and research hypotheses.*

    The null hypothesis, shown below, states that the sequence of events is random. The research hypothesis, shown below, states that the sequence of events is not random.

    The null hypothesis is

    H_0: The sequence in which the teacher calls on males and females is random.

    The research hypothesis is

    H_A: The sequence in which the teacher calls on males and females is not random.

THE RUNS TEST FOR RANDOMNESS

2. *Set the level of risk (or the level of significance) associated with the null hypothesis.*

   The level of risk, also called an alpha (α), is frequently set at 0.05. We will use an alpha of 0.05 in our example. In other words, there is a 95% chance that any observed statistical difference will be real and not due to chance.

3. *Choose the appropriate test statistic.*

   The observer is examining the data for randomness. Therefore, he is using a runs test for randomness.

4. *Compute the test statistic.*

   First, determine the number of runs, R. It is helpful to separate the events as shown in Figure 9.11. The number of runs in the sequence is $R = 17$.

$$
\begin{array}{cccccccccc}
\text{FF} & \text{MM} & \text{FFF} & \text{M} & \text{FFF} & \text{MM} & \text{F} & \text{MMMM} & \text{F} \\
1 & 2 & 3 & 4 & 5 & 6 & 7 & 8 & 9
\end{array}
$$

$$
\begin{array}{ccccccc}
\text{MMMMMM} & \text{F} & \text{M} & \text{F} & \text{MM} & \text{F} & \text{MMMMM} & \text{F} \\
10 & 11 & 12 & 13 & 14 & 15 & 16 & 17
\end{array}
$$

FIGURE 9.11

Our number of values exceeds those available from our critical values table for the runs test (Table B.10 is limited to $n_1 \leq 20$ and $n_2 \leq 20$). Therefore, we will find a z-score for our data using a normal approximation. First, we must find the mean, $\bar{x}_R$, and the standard deviation, s_R, for the data.

$$\bar{x}_R = \frac{2n_1 n_2}{n_1 + n_2} + 1 = \frac{2(23)(14)}{23 + 14} + 1$$

$$= \frac{644}{37} + 1 = 17.4 + 1$$

$$= 18.4$$

and

$$s_R = \sqrt{\frac{2n_1 n_2 (2n_1 n_2 - n_1 - n_2)}{(n_1 + n_2)^2 (n_1 + n_2 - 1)}} = \sqrt{\frac{2(23)(14)(2(23)(14) - 23 - 14)}{(23 + 14)^2 (23 + 14 - 1)}}$$

$$= \sqrt{\frac{(644)(644 - 23 - 14)}{(37)^2 (36)}} = \sqrt{\frac{(644)(607)}{(1369)(36)}}$$

$$= \sqrt{\frac{390908}{49284}} = \sqrt{7.932}$$

$$= 2.816$$

Next, we calculate a z-score. We use the correction for continuity, mean, standard deviation, and number of runs ($R = 17$) to calculate a z-score. The only value that we still need is the correction for continuity, h. Recall that $h = +0.5$ if $R < (2n_1n_2/(n_1 + n_2) + 1)$, and $h = -0.5$ if $R > (2n_1n_2/(n_1 + n_2) + 1)$. In our example, $2n_1n_2/(n_1 + n_2) + 1 = 2(23)(14)/(23 + 14) + 1 = 18.4$. Since $17 < 18.4$, we choose $h = +0.5$.

Now, we use our z-score formula with correction for continuity.

$$z^* = \frac{R + h - \bar{x}_R}{s_R} = \frac{17 + (+0.5) - 18.4}{2.816}$$
$$= -0.3196$$

5. *Determine the value needed for rejection of the null hypothesis using the appropriate table of critical values for the particular statistic.*

    Table B.1 is used to establish the critical region of z-scores. For a two-tailed test with $\alpha = 0.05$, we must not reject the null hypothesis if $-1.96 \leq z^* \leq 1.96$.

6. *Compare the obtained value to the critical value.*

    We find that z^* is within the critical region of the distribution, $-1.96 \leq -0.3196 \leq 1.96$. Therefore, we do not reject the null hypothesis. This suggests that the order in which the science teacher calls on males and females is random.

7. *Interpret the results.*

    We did not reject the null hypothesis, suggesting that the sequence of events is random. Therefore, our data indicate that the order in which the science teacher calls on males and females is random.

8. *Reporting the results.*

    Based on our analysis, the runs test indicated that the sequence was random ($R = 17, n_1 = 23, n_2 = 14, p > 0.05$). Therefore, the study provides evidence that the science teacher was demonstrating no gender bias.

9.3.4 Sample Runs Test Referencing a Custom Value

The science teacher in the earlier example wishes to examine the pattern of an "at-risk" student's weekly quiz performance. A passing quiz score is 70. Sometimes the student failed and other times he passed. The teacher wished to determine if the student's performance is random or not. Table 9.1 shows the student's weekly quiz scores for a 12-week period.

1. *State the null and research hypotheses.*

    The null hypothesis, shown below, states that the sequence of events is random. The research hypothesis, shown below, states that the sequence of events is not random.

THE RUNS TEST FOR RANDOMNESS

TABLE 9.1

Week	Student Quiz Scores
1	65
2	55
3	95
4	15
5	75
6	65
7	80
8	75
9	60
10	55
11	75
12	80

The null hypothesis is

H_0: The sequence in which the student passes and fails a weekly science quiz is random.

The research hypothesis is

H_A: The sequence in which the student passes and fails a weekly science quiz is not random.

2. *Set the level of risk (or the level of significance) associated with the null hypothesis.*

   The level of risk, also called an alpha (α), is frequently set at 0.05. We will use an alpha of 0.05 in our example. In other words, there is a 95% chance that any observed statistical difference will be real and not due to chance.

3. *Choose the appropriate test statistic.*

   The observer is examining the data for randomness. Therefore, he is using a runs test for randomness.

4. *Compute the test statistic.*

   The custom value is 69.9 since a passing quiz score is 70. We must identify which quiz scores fall above the custom score and which quiz scores fall below it. As shown in Table 9.2, we mark the quiz scores that fall above the custom score with $+$ and the quiz scores that fall below with $-$. Then, we count the number of runs, R. The number of runs in the above sequence is $R = 8$.

5. *Determine the value needed for rejection of the null hypothesis using the appropriate table of critical values for the particular statistic.*

   Since the sample sizes are small, we refer to Table B.10, which lists the critical values for the runs test. The critical values are found in the table at the point for $n_1 = 6$ and $n_2 = 6$. We set $\alpha = 0.05$. The critical region for the runs test

TABLE 9.2

Student	Week Quiz Scores	Relation to Custom Score
1	65	−
2	55	−
3	95	+
4	15	−
5	75	+
6	65	−
7	80	+
8	75	+
9	60	−
10	55	−
11	70	+
12	80	+

is $3 < R < 11$. If the number of runs, R, is 3 or less or 11 or greater, we reject our null hypothesis.

6. *Compare the obtained value to the critical value.*

    We found that $R = 8$. This value is within our critical region ($3 < R < 11$). Therefore, we must not reject the null hypothesis.

7. *Interpret the results.*

    We did not reject the null hypothesis, suggesting that the sequence of events is random. Therefore, we can state that based on a passing score of 70, the student's weekly science quiz performance is random.

8. *Reporting the results.*

    For this example, the runs test indicated that the sequence was random ($R = 8$, $n_1 = 6$, $n_2 = 6$, $p > 0.05$). Therefore, the evidence suggests that the pattern of the student's weekly science quiz performance is random in terms of achieving a passing score of 70.

9.3.5 Performing the Runs Test for a Custom Value Using SPSS

We will analyze the data from the above example using SPSS.

1. *Define your variables.*

    First, click the "Variable View" tab at the bottom of your screen. As shown in Figure 9.12, type the names of your variables in the "Name" column. We call our variable "Quiz".

2. *Type in your values.*

    Click the "Data View" tab at the bottom of your screen, as shown in Figure 9.13. Type the values into the column in the same order they occurred.

THE RUNS TEST FOR RANDOMNESS

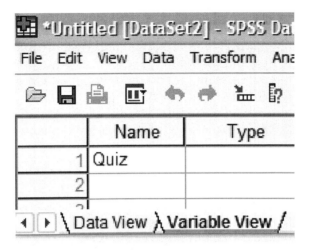

FIGURE 9.12

	Quiz
1	65.00
2	55.00
3	95.00
4	15.00
5	75.00
6	65.00
7	80.00
8	75.00
9	60.00
10	55.00
11	70.00
12	80.00

FIGURE 9.13

TEST FOR RANDOMNESS: THE RUNS TEST

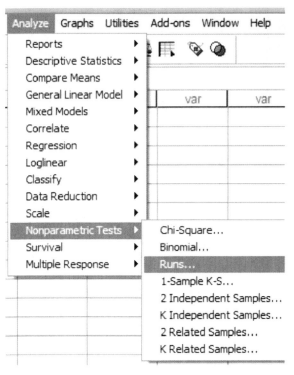

FIGURE 9.14

3. *Analyze your data.*

As shown in Figure 9.14, use the pull-down menus to choose "Analyze", "Nonparametric Tests", and "Runs...".

As shown in Figure 9.15 under "Cut Point", uncheck "Median" and check the box next to "Custom:" Type the custom value in the box. For our example,

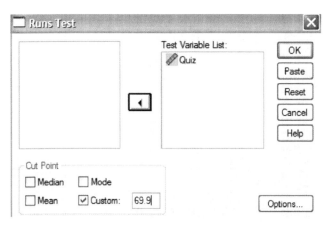

FIGURE 9.15

we use a custom value of 69.9. Next, use the arrow button to place the variable with your data values in the box labeled "Test Variable List". In our example, we chose the variable "Quiz".

4. *Interpret the results from the SPSS Output window.*

Runs Test

	Quiz
Test Value[a]	69.9000
Total Cases	12
Number of Runs	8
Z	.303
Asymp. Sig. (2-tailed)	.762

a. User-specified.

The runs test SPSS output table returns the custom test value (69.9), the total number of observations ($N = 12$), and the number of runs ($R = 8$). SPSS also calculates the z-score ($z^* = 0.303$) and the two-tailed significance ($p = 0.762$).

5. *Determine the observation frequencies for each event.*

In order to determine the number of observations for each event, an additional set of steps is required. As shown in Figure 9.16, use the pull-down menus to choose "Analyze", "Descriptive Statistics", and "Frequencies...". Next, use the arrow button to place the variable with your data values in the box labeled "Variable(s):", as shown in Figure 9.17. In our example, we chose the variable "Quiz". Finally, click "OK" to perform the analysis.

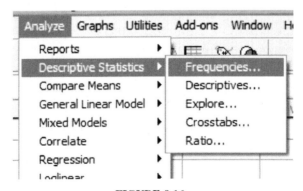

FIGURE 9.16

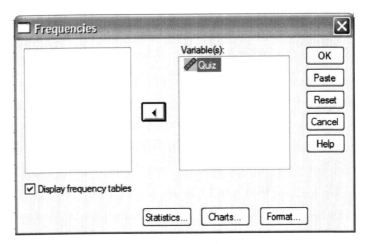

FIGURE 9.17

Quiz

		Frequency	Percent	Valid Percent	Cumulative Percent
Valid	15.00	1	8.3	8.3	8.3
	55.00	2	16.7	16.7	25.0
	60.00	1	8.3	8.3	33.3
	65.00	2	16.7	16.7	50.0
	70.00	1	8.3	8.3	58.3
	75.00	2	16.7	16.7	75.0
	80.00	2	16.7	16.7	91.7
	95.00	1	8.3	8.3	100.0
	Total	12	100.0	100.0	

The second SPSS output table displays the frequencies for each value. You must count the number of values above the custom value and the number values below it to determine the frequency for each event.

Based on the results from SPSS, the runs test indicated that the sequence was random ($R = 8, n_1 = 6, n_2 = 6, p > 0.05$). Therefore, the pattern of the student's weekly science quiz performance is random in terms of achieving a passing score of 70.

9.4 EXAMPLES FROM THE LITERATURE

Below are varied examples of the nonparametric procedures described in this chapter. We have summarized each study's research problem and researchers' rationale(s) for choosing a nonparametric approach. We encourage you to obtain these studies if you are interested in their results.

- Dorsey-Palmateer, R. & Smith, G. (2004) Bowlers' hot hands. *The American Statistician, 58*(1), 38–45.

  Dorsey-Palmateer and Smith call into question a classical statistics experiment that debunked a commonly held belief that basketball players' shooting accuracy is based on the performance immediately preceding a given shot. The authors explore this notion of hot hands among professional bowlers. They examine a series of rolls and differentiate between strikes and nonstrikes. They used a runs test to analyze the sequence of bowlers' performance for randomness.

- Vergin, R. C. (2000). Winning streaks in sports and the misperception of momentum. *Journal of Sport Behavior, 23*(2), 181–197.

  Vergin explored the presence of momentum among Major League Baseball (MLB) teams and National Basketball Association (NBA) teams. He described momentum as the tendency for a winning team to continue to win and a losing team to continue to lose. Therefore, he used a Wald–Wolfowitz runs test to examine the winning and losing streaks of the 28 MLB teams in 1996 and of the 29 NBA teams during the 1996–1997 and 1997–1998 seasons.

- Pollay, R. W., Lee, J. S., & Carter-Whitney, D. (1992). Separate, but not equal: Racial segmentation in cigarette advertising. *Journal of Advertising, 21*(1), 45–57.

  Pollay, Lee, and Carter-Whitney investigated the possibility that cigarette companies segregated and segmented advertising efforts toward black consumers. They used a runs test to compare the change in annual frequency of cigarette ads that appeared in Life Magazine versus Ebony.

9.5 SUMMARY

The runs test is a statistical procedure for examining a series of events for randomness. This nonparametric test has no parametric equivalent. In this chapter, we described how to perform and interpret a runs test for both small samples and large samples. We also explained how to perform the procedure using SPSS. Finally, we offered varied examples of these nonparametric statistics from the literature.

9.6 PRACTICE QUESTIONS

1. The data below represent the daily performance of a popular stock. Letter A represents a gain and letter B represents a loss. Use a runs test to analyze the stock's performance for randomness. Set $\alpha = 0.05$. Report the results.

   BAABBAABBBBBAABAAAAB

2. A machine on an automated assembly line produces a unique type of bolt. If the machine fails more than three times in an hour, the total production on the line is slowed down. The machine has often exceeded the number of acceptable failures for the last week. The machine is expensive and more cost effective to repair, but the maintenance crew cannot find the problem. The

TABLE 9.3

Hour	Number of Failures
1	6
2	4
3	2
4	2
5	7
6	5
7	7
8	9
9	2
10	0
11	0
12	0
13	7
14	6
15	5
16	9
17	1
18	0
19	1
20	8
21	5
22	9
23	4
24	5

plant manager asks you to determine if the failure rates are random or if a pattern exists. Table 9.3 shows the number of failures per hour for a 24 hour period.

Use a runs test with a custom value of 3.1 to analyze the acceptable/unacceptable failure rate for randomness. Set $\alpha = 0.05$. Report the results.

9.7 SOLUTIONS TO PRACTICE QUESTIONS

1. The results from the analysis are displayed in the SPSS Outputs below.

Runs Test

	Performance
Test Value[a]	1.0000
Total Cases	20
Number of Runs	9
Z	-.689
Asymp. Sig. (2-tailed)	.491

a. User-specified.

SOLUTIONS TO PRACTICE QUESTIONS

Performance

		Frequency	Percent	Valid Percent	Cumulative Percent
Valid	gain	10	47.6	50.0	50.0
	loss	10	47.6	50.0	100.0
	Total	20	95.2	100.0	
Missing	System	1	4.8		
Total		21	100.0		

The sequence of the stock's gains and losses was random ($R = 9$, $n_1 = 10$, $n_2 = 10$, $p > 0.05$).

2. The results from the analysis are displayed in the SPSS Outputs below.

Runs Test

	Failure_Rate
Test Value[a]	3.1000
Total Cases	24
Number of Runs	7
Z	-2.121
Asymp. Sig. (2-tailed)	.034

a. User-specified.

Failure_Rate

		Frequency	Percent	Valid Percent	Cumulative Percent
Valid	.00	4	16.7	16.7	16.7
	1.00	2	8.3	8.3	25.0
	2.00	3	12.5	12.5	37.5
	4.00	2	8.3	8.3	45.8
	5.00	4	16.7	16.7	62.5
	6.00	2	8.3	8.3	70.8
	7.00	3	12.5	12.5	83.3
	8.00	1	4.2	4.2	87.5
	9.00	3	12.5	12.5	100.0
	Total	24	100.0	100.0	

The sequence of the machine's acceptable/unacceptable failure rate was not random ($R = 7$, $n_1 = 9$, $n_2 = 15$, $p < 0.05$).

APPENDIX A

SPSS AT A GLANCE

A.1 INTRODUCTION

Statistical Package for Social Sciences, or SPSS, is a powerful tool for performing statistical analyses. Once you learn some basics, you will be able to save hours of algebraic computations while producing meaningful results. This section of the appendices includes a very basic overview of SPSS. We recommend you also run the tutorial when the program initially starts. The tutorial offers you a more detailed description of how to use the program.

A.2 OPENING SPSS

Begin by launching SPSS like any normal application. After SPSS begins, a window will appear, as seen in Figure A.1. Choose "Type in data" and click OK. From this screen, you can also view the SPSS tutorial or open a file with existing data.

A.3 INPUTTING DATA

The SPSS Data Editor window is shown in Figure A.2a. This window will allow you to type in your data. Notice the "Data View" and the "Variable View" tabs at the bottom of the window. Before inputting values, we must set up our variables.

Nonparametric Statistics for Non-Statisticians, Gregory W. Corder and Dale I. Foreman
Copyright © 2009 John Wiley & Sons, Inc.

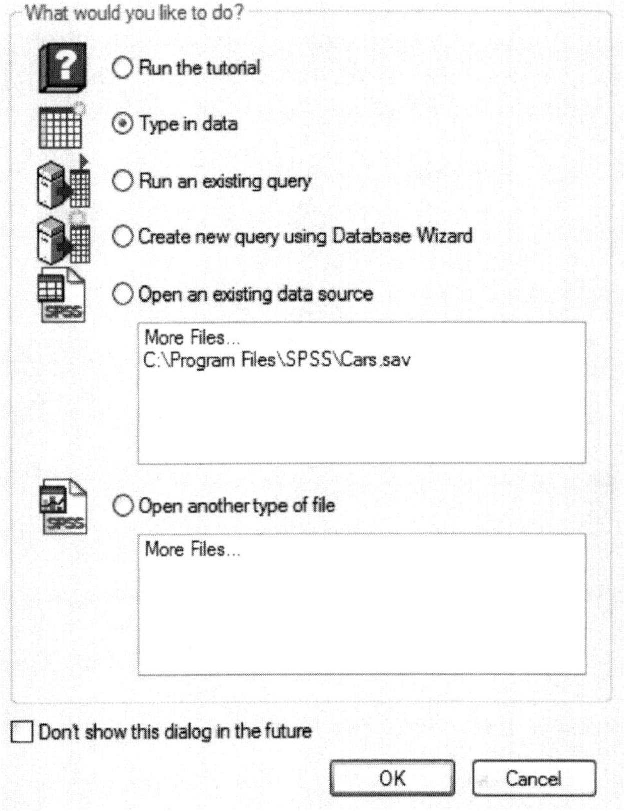

APPENDIX A: SPSS AT A GLANCE

What would you like to do?

- ○ Run the tutorial
- ⊙ Type in data
- ○ Run an existing query
- ○ Create new query using Database Wizard
- ○ Open an existing data source

  More Files...
  C:\Program Files\SPSS\Cars.sav

- ○ Open another type of file

  More Files...

☐ Don't show this dialog in the future

[OK]   [Cancel]

FIGURE A.1

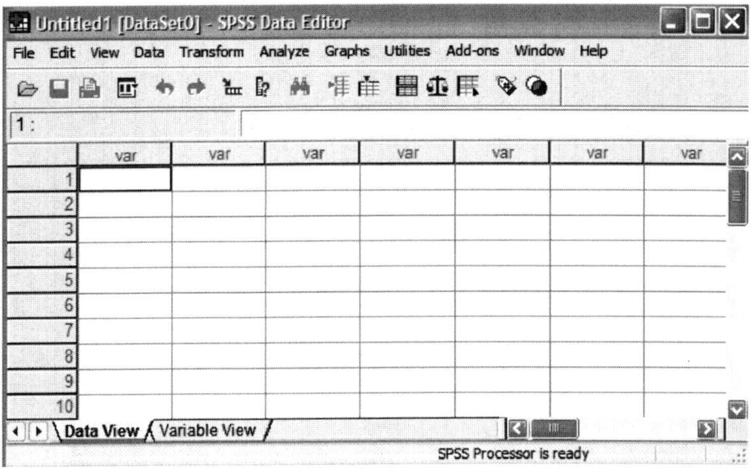

FIGURE A.2a

1. Select the "Variable View" tab located at the bottom of the SPSS Data Editor window to define the characteristics of your variables.
2. Once you change the window to Variable View, as seen in Figure A.2b, type in the names for each variable in the "Names" field. SPSS will not accept spaces at this step, so use underscores. For example, use "Test_A" instead of "Test A".

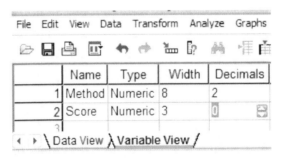

FIGURE A.2b

3. In the "Width" field, choose the maximum number of characters for each value.
4. In the "Decimals" field, choose the number of decimals for each value. For the "Score" variable, we have changed the width to 3 characters and the decimals to 0, as seen in Figure A.2c.

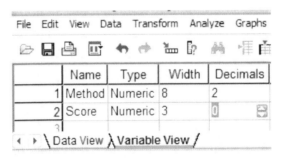

FIGURE A.2c

5. Use the "Label" field to assign names to the variables. Those names will appear in the output report that SPSS returns after an analysis. In addition, the "Label" field will allow you to use spaces, unlike the "Name" field. Figure A.2d illustrates that "Teaching Method" will identify the "Method" variable in the SPSS output report.

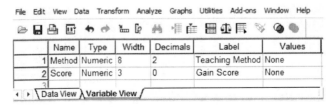

FIGURE A.2d

APPENDIX A: SPSS AT A GLANCE   215

6. Use the "Values" field to assign a value to categorical data. As shown in Figure A.2e, clicking on the gray box in the right side of the column will cause a new window to appear that allows you to input your settings. As seen in Figure A.2e, the "One-on-One" teaching method is assigned a value of 1 and the "Small Group" teaching method is assigned a value of 2.

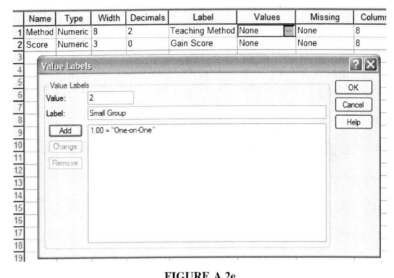

FIGURE A.2e

7. The "Align" field allows you to change the alignment of the values as they appear in the "Data view" tab.
8. The "Measure" field allows you to select the type of scale. In Figure A.2f, we are preparing to change the "Method" variable from a scale to a nominal measure.

FIGURE A.2f

9. Finally, click on the "Data view" tab at the bottom of the screen and manually type in your data values or paste them from a spreadsheet. Figure A.2g shows the values entered into the Data Editor.

	Method	Score
1	1.00	16
2	1.00	13
3	1.00	16
4	1.00	16
5	1.00	13
6	1.00	9
7	1.00	12
8	1.00	12
9	1.00	20
10	1.00	17
11	2.00	11
12	2.00	2
13	2.00	10
14	2.00	4
15	2.00	9
16	2.00	8
17	2.00	5
18	2.00	6
19	2.00	4
20	2.00	16
21		

\Data View ⟨ Variable View /

FIGURE A.2g

A.4 ANALYZING DATA

Choose "Analyze" from the pull-down menus at the top of the Data Editor window and select the appropriate test. In the figure below, notice that "Nonparametric Tests" has been selected. This menu presents all of the tests discussed in this book.

If your menu choices look slightly different from Figure A.3, it may be that you are using the student version of SPSS. The student version is far more powerful than most people will ever need, and even though the figures in this book are from the full version of SPSS, we doubt that you will notice any difference.

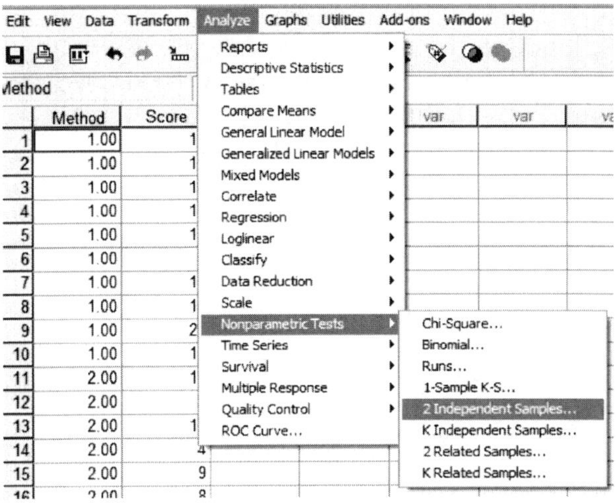

APPENDIX A: SPSS AT A GLANCE

FIGURE A.3

A.5 THE SPSS OUTPUT

Once SPSS has performed an analysis, a new window will appear called the "SPSS Viewer", as seen in Figure A.4. The window is separated into two panes. The left pane

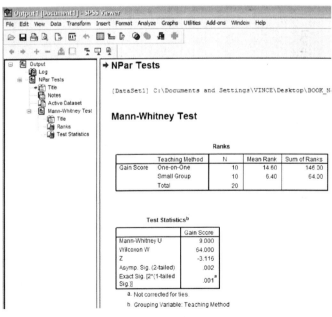

FIGURE A.4

is called the outline pane and shows all of the information stored in the viewer. The pane to the right is called the contents pane and shows the actual output of the analysis.

You may wish to include any tables or graphs from the output pane in a report. You can select the objects in the contents pane and copy them onto a word processing document. The small arrow identifies the selected object.

APPENDIX B

TABLES OF CRITICAL VALUES

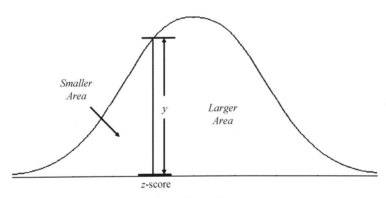

FIGURE B.1

TABLE B.1 The Normal Distribution

z-Score	Smaller Area	Larger Area	y
0.00	0.5000	0.5000	0.3989
0.01	0.4960	0.5040	0.3989
0.02	0.4920	0.5080	0.3989
			(Continued)

Nonparametric Statistics for Non-Statisticians, Gregory W. Corder and Dale I. Foreman
Copyright © 2009 John Wiley & Sons, Inc.

TABLE B.1 (*Continued*)

z-Score	Smaller Area	Larger Area	y
0.03	0.4880	0.5120	0.3988
0.04	0.4840	0.5160	0.3986
0.05	0.4801	0.5199	0.3984
0.06	0.4761	0.5239	0.3982
0.07	0.4721	0.5279	0.3980
0.08	0.4681	0.5319	0.3977
0.09	0.4641	0.5359	0.3973
0.10	0.4602	0.5398	0.3970
0.11	0.4562	0.5438	0.3965
0.12	0.4522	0.5478	0.3961
0.13	0.4483	0.5517	0.3956
0.14	0.4443	0.5557	0.3951
0.15	0.4404	0.5596	0.3945
0.16	0.4364	0.5636	0.3939
0.17	0.4325	0.5675	0.3932
0.18	0.4286	0.5714	0.3925
0.19	0.4247	0.5753	0.3918
0.20	0.4207	0.5793	0.3910
0.21	0.4168	0.5832	0.3902
0.22	0.4129	0.5871	0.3894
0.23	0.4090	0.5910	0.3885
0.24	0.4052	0.5948	0.3876
0.25	0.4013	0.5987	0.3867
0.26	0.3974	0.6026	0.3857
0.27	0.3936	0.6064	0.3847
0.28	0.3897	0.6103	0.3836
0.29	0.3859	0.6141	0.3825
0.30	0.3821	0.6179	0.3814
0.31	0.3783	0.6217	0.3802
0.32	0.3745	0.6255	0.3790
0.33	0.3707	0.6293	0.3778
0.34	0.3669	0.6331	0.3765
0.35	0.3632	0.6368	0.3752
0.36	0.3594	0.6406	0.3739
0.37	0.3557	0.6443	0.3725
0.38	0.3520	0.6480	0.3712
0.39	0.3483	0.6517	0.3697
0.40	0.3446	0.6554	0.3683
0.41	0.3409	0.6591	0.3668
0.42	0.3372	0.6628	0.3653
0.43	0.3336	0.6664	0.3637
0.44	0.3300	0.6700	0.3621
0.45	0.3264	0.6736	0.3605
0.46	0.3228	0.6772	0.3589
0.47	0.3192	0.6808	0.3572
0.48	0.3156	0.6844	0.3555

TABLE B.1 (*Continued*)

z-Score	Smaller Area	Larger Area	y
0.49	0.3121	0.6879	0.3538
0.50	0.3085	0.6915	0.3521
0.51	0.3050	0.6950	0.3503
0.52	0.3015	0.6985	0.3485
0.53	0.2981	0.7019	0.3467
0.54	0.2946	0.7054	0.3448
0.55	0.2912	0.7088	0.3429
0.56	0.2877	0.7123	0.3410
0.57	0.2843	0.7157	0.3391
0.58	0.2810	0.7190	0.3372
0.59	0.2776	0.7224	0.3352
0.60	0.2743	0.7257	0.3332
0.61	0.2709	0.7291	0.3312
0.62	0.2676	0.7324	0.3292
0.63	0.2643	0.7357	0.3271
0.64	0.2611	0.7389	0.3251
0.65	0.2578	0.7422	0.3230
0.66	0.2546	0.7454	0.3209
0.67	0.2514	0.7486	0.3187
0.68	0.2483	0.7517	0.3166
0.69	0.2451	0.7549	0.3144
0.70	0.2420	0.7580	0.3123
0.71	0.2389	0.7611	0.3101
0.72	0.2358	0.7642	0.3079
0.73	0.2327	0.7673	0.3056
0.74	0.2296	0.7704	0.3034
0.75	0.2266	0.7734	0.3011
0.76	0.2236	0.7764	0.2989
0.77	0.2206	0.7794	0.2966
0.78	0.2177	0.7823	0.2943
0.79	0.2148	0.7852	0.2920
0.80	0.2119	0.7881	0.2897
0.81	0.2090	0.7910	0.2874
0.82	0.2061	0.7939	0.2850
0.83	0.2033	0.7967	0.2827
0.84	0.2005	0.7995	0.2803
0.85	0.1977	0.8023	0.2780
0.86	0.1949	0.8051	0.2756
0.87	0.1922	0.8078	0.2732
0.88	0.1894	0.8106	0.2709
0.89	0.1867	0.8133	0.2685
0.90	0.1841	0.8159	0.2661
0.91	0.1814	0.8186	0.2637
0.92	0.1788	0.8212	0.2613
0.93	0.1762	0.8238	0.2589
0.94	0.1736	0.8264	0.2565

(*Continued*)

TABLE B.1 (*Continued*)

z-Score	Smaller Area	Larger Area	y
0.95	0.1711	0.8289	0.2541
0.96	0.1685	0.8315	0.2516
0.97	0.1660	0.8340	0.2492
0.98	0.1635	0.8365	0.2468
0.99	0.1611	0.8389	0.2444
1.00	0.1587	0.8413	0.2420
1.01	0.1562	0.8438	0.2396
1.02	0.1539	0.8461	0.2371
1.03	0.1515	0.8485	0.2347
1.04	0.1492	0.8508	0.2323
1.05	0.1469	0.8531	0.2299
1.06	0.1446	0.8554	0.2275
1.07	0.1423	0.8577	0.2251
1.08	0.1401	0.8599	0.2227
1.09	0.1379	0.8621	0.2203
1.10	0.1357	0.8643	0.2179
1.11	0.1335	0.8665	0.2155
1.12	0.1314	0.8686	0.2131
1.13	0.1292	0.8708	0.2107
1.14	0.1271	0.8729	0.2083
1.15	0.1251	0.8749	0.2059
1.16	0.1230	0.8770	0.2036
1.17	0.1210	0.8790	0.2012
1.18	0.1190	0.8810	0.1989
1.19	0.1170	0.8830	0.1965
1.20	0.1151	0.8849	0.1942
1.21	0.1131	0.8869	0.1919
1.22	0.1112	0.8888	0.1895
1.23	0.1093	0.8907	0.1872
1.24	0.1075	0.8925	0.1849
1.25	0.1056	0.8944	0.1826
1.26	0.1038	0.8962	0.1804
1.27	0.1020	0.8980	0.1781
1.28	0.1003	0.8997	0.1758
1.29	0.0985	0.9015	0.1736
1.30	0.0968	0.9032	0.1714
1.31	0.0951	0.9049	0.1691
1.32	0.0934	0.9066	0.1669
1.33	0.0918	0.9082	0.1647
1.34	0.0901	0.9099	0.1626
1.35	0.0885	0.9115	0.1604
1.36	0.0869	0.9131	0.1582
1.37	0.0853	0.9147	0.1561
1.38	0.0838	0.9162	0.1539
1.39	0.0823	0.9177	0.1518
1.40	0.0808	0.9192	0.1497

TABLE B.1 (*Continued*)

z-Score	Smaller Area	Larger Area	y
1.41	0.0793	0.9207	0.1476
1.42	0.0778	0.9222	0.1456
1.43	0.0764	0.9236	0.1435
1.44	0.0749	0.9251	0.1415
1.45	0.0735	0.9265	0.1394
1.46	0.0721	0.9279	0.1374
1.47	0.0708	0.9292	0.1354
1.48	0.0694	0.9306	0.1334
1.49	0.0681	0.9319	0.1315
1.50	0.0668	0.9332	0.1295
1.51	0.0655	0.9345	0.1276
1.52	0.0643	0.9357	0.1257
1.53	0.0630	0.9370	0.1238
1.54	0.0618	0.9382	0.1219
1.55	0.0606	0.9394	0.1200
1.56	0.0594	0.9406	0.1182
1.57	0.0582	0.9418	0.1163
1.58	0.0571	0.9429	0.1145
1.59	0.0559	0.9441	0.1127
1.60	0.0548	0.9452	0.1109
1.61	0.0537	0.9463	0.1092
1.62	0.0526	0.9474	0.1074
1.63	0.0516	0.9484	0.1057
1.64	0.0505	0.9495	0.1040
1.65	0.0495	0.9505	0.1023
1.66	0.0485	0.9515	0.1006
1.67	0.0475	0.9525	0.0989
1.68	0.0465	0.9535	0.0973
1.69	0.0455	0.9545	0.0957
1.70	0.0446	0.9554	0.0940
1.71	0.0436	0.9564	0.0925
1.72	0.0427	0.9573	0.0909
1.73	0.0418	0.9582	0.0893
1.74	0.0409	0.9591	0.0878
1.75	0.0401	0.9599	0.0863
1.76	0.0392	0.9608	0.0848
1.77	0.0384	0.9616	0.0833
1.78	0.0375	0.9625	0.0818
1.79	0.0367	0.9633	0.0804
1.80	0.0359	0.9641	0.0790
1.81	0.0351	0.9649	0.0775
1.82	0.0344	0.9656	0.0761
1.83	0.0336	0.9664	0.0748
1.84	0.0329	0.9671	0.0734
1.85	0.0322	0.9678	0.0721

(*Continued*)

TABLE B.1 (*Continued*)

z-Score	Smaller Area	Larger Area	y
1.86	0.0314	0.9686	0.0707
1.87	0.0307	0.9693	0.0694
1.88	0.0301	0.9699	0.0681
1.89	0.0294	0.9706	0.0669
1.90	0.0287	0.9713	0.0656
1.91	0.0281	0.9719	0.0644
1.92	0.0274	0.9726	0.0632
1.93	0.0268	0.9732	0.0620
1.94	0.0262	0.9738	0.0608
1.95	0.0256	0.9744	0.0596
1.96	0.0250	0.9750	0.0584
1.97	0.0244	0.9756	0.0573
1.98	0.0239	0.9761	0.0562
1.99	0.0233	0.9767	0.0551
2.00	0.0228	0.9772	0.0540
2.01	0.0222	0.9778	0.0529
2.02	0.0217	0.9783	0.0519
2.03	0.0212	0.9788	0.0508
2.04	0.0207	0.9793	0.0498
2.05	0.0202	0.9798	0.0488
2.06	0.0197	0.9803	0.0478
2.07	0.0192	0.9808	0.0468
2.08	0.0188	0.9812	0.0459
2.09	0.0183	0.9817	0.0449
2.10	0.0179	0.9821	0.0440
2.11	0.0174	0.9826	0.0431
2.12	0.0170	0.9830	0.0422
2.13	0.0166	0.9834	0.0413
2.14	0.0162	0.9838	0.0404
2.15	0.0158	0.9842	0.0396
2.16	0.0154	0.9846	0.0387
2.17	0.0150	0.9850	0.0379
2.18	0.0146	0.9854	0.0371
2.19	0.0143	0.9857	0.0363
2.20	0.0139	0.9861	0.0355
2.21	0.0136	0.9864	0.0347
2.22	0.0132	0.9868	0.0339
2.23	0.0129	0.9871	0.0332
2.24	0.0125	0.9875	0.0325
2.25	0.0122	0.9878	0.0317
2.26	0.0119	0.9881	0.0310
2.27	0.0116	0.9884	0.0303
2.28	0.0113	0.9887	0.0297
2.29	0.0110	0.9890	0.0290
2.30	0.0107	0.9893	0.0283
2.31	0.0104	0.9896	0.0277

TABLE B.1 (*Continued*)

z-Score	Smaller Area	Larger Area	y
2.32	0.0102	0.9898	0.0270
2.33	0.0099	0.9901	0.0264
2.34	0.0096	0.9904	0.0258
2.35	0.0094	0.9906	0.0252
2.36	0.0091	0.9909	0.0246
2.37	0.0089	0.9911	0.0241
2.38	0.0087	0.9913	0.0235
2.39	0.0084	0.9916	0.0229
2.40	0.0082	0.9918	0.0224
2.41	0.0080	0.9920	0.0219
2.42	0.0078	0.9922	0.0213
2.43	0.0075	0.9925	0.0208
2.44	0.0073	0.9927	0.0203
2.45	0.0071	0.9929	0.0198
2.46	0.0069	0.9931	0.0194
2.47	0.0068	0.9932	0.0189
2.48	0.0066	0.9934	0.0184
2.49	0.0064	0.9936	0.0180
2.50	0.0062	0.9938	0.0175
2.51	0.0060	0.9940	0.0171
2.52	0.0059	0.9941	0.0167
2.53	0.0057	0.9943	0.0163
2.54	0.0055	0.9945	0.0158
2.55	0.0054	0.9946	0.0154
2.56	0.0052	0.9948	0.0151
2.57	0.0051	0.9949	0.0147
2.58	0.0049	0.9951	0.0143
2.59	0.0048	0.9952	0.0139
2.60	0.0047	0.9953	0.0136
2.61	0.0045	0.9955	0.0132
2.62	0.0044	0.9956	0.0129
2.63	0.0043	0.9957	0.0126
2.64	0.0041	0.9959	0.0122
2.65	0.0040	0.9960	0.0119
2.66	0.0039	0.9961	0.0116
2.67	0.0038	0.9962	0.0113
2.68	0.0037	0.9963	0.0110
2.69	0.0036	0.9964	0.0107
2.70	0.0035	0.9965	0.0104
2.71	0.0034	0.9966	0.0101
2.72	0.0033	0.9967	0.0099
2.73	0.0032	0.9968	0.0096
2.74	0.0031	0.9969	0.0093
2.75	0.0030	0.9970	0.0091
2.76	0.0029	0.9971	0.0088

(*Continued*)

TABLE B.1 (*Continued*)

z-Score	Smaller Area	Larger Area	y
2.77	0.0028	0.9972	0.0086
2.78	0.0027	0.9973	0.0084
2.79	0.0026	0.9974	0.0081
2.80	0.0026	0.9974	0.0079
2.81	0.0025	0.9975	0.0077
2.82	0.0024	0.9976	0.0075
2.83	0.0023	0.9977	0.0073
2.84	0.0023	0.9977	0.0071
2.85	0.0022	0.9978	0.0069
2.86	0.0021	0.9979	0.0067
2.87	0.0021	0.9979	0.0065
2.88	0.0020	0.9980	0.0063
2.89	0.0019	0.9981	0.0061
2.90	0.0019	0.9981	0.0060
2.91	0.0018	0.9982	0.0058
2.92	0.0018	0.9982	0.0056
2.93	0.0017	0.9983	0.0055
2.94	0.0016	0.9984	0.0053
2.95	0.0016	0.9984	0.0051
2.96	0.0015	0.9985	0.0050
2.97	0.0015	0.9985	0.0048
2.98	0.0014	0.9986	0.0047
2.99	0.0014	0.9986	0.0046
3.00	0.0013	0.9987	0.0044
3.10	0.0010	0.9990	0.0033
3.20	0.0007	0.9993	0.0024
3.30	0.0005	0.9995	0.0017
3.50	0.0002	0.9998	0.0009
3.75	0.0001	0.9999	0.0004
4.00	0.0000	1.0000	0.0001

TABLE B.2 The Chi-Square Distribution

df	0.99	0.975	0.95	0.9	0.1	0.05	0.025	0.01
1	0.00	0.00	0.00	0.02	2.71	3.84	5.02	6.63
2	0.02	0.05	0.10	0.21	4.61	5.99	7.38	9.21
3	0.11	0.22	0.35	0.58	6.25	7.81	9.35	11.34
4	0.30	0.48	0.71	1.06	7.78	9.49	11.14	13.28
5	0.55	0.83	1.15	1.61	9.24	11.07	12.83	15.09
6	0.87	1.24	1.64	2.20	10.64	12.59	14.45	16.81
7	1.24	1.69	2.17	2.83	12.02	14.07	16.01	18.48
8	1.65	2.18	2.73	3.49	13.36	15.51	17.53	20.09
9	2.09	2.70	3.33	4.17	14.68	16.92	19.02	21.67
10	2.56	3.25	3.94	4.87	15.99	18.31	20.48	23.21
11	3.05	3.82	4.57	5.58	17.28	19.68	21.92	24.73

APPENDIX B: TABLES OF CRITICAL VALUES

TABLE B.2 (*Continued*)

df	0.99	0.975	0.95	0.9	0.1	0.05	0.025	0.01
12	3.57	4.40	5.23	6.30	18.55	21.03	23.34	26.22
13	4.11	5.01	5.89	7.04	19.81	22.36	24.74	27.69
14	4.66	5.63	6.57	7.79	21.06	23.68	26.12	29.14
15	5.23	6.26	7.26	8.55	22.31	25.00	27.49	30.58
16	5.81	6.91	7.96	9.31	23.54	26.30	28.85	32.00
17	6.41	7.56	8.67	10.09	24.77	27.59	30.19	33.41
18	7.01	8.23	9.39	10.86	25.99	28.87	31.53	34.81
19	7.63	8.91	10.12	11.65	27.20	30.14	32.85	36.19
20	8.26	9.59	10.85	12.44	28.41	31.41	34.17	37.57
21	8.90	10.28	11.59	13.24	29.62	32.67	35.48	38.93
22	9.54	10.98	12.34	14.04	30.81	33.92	36.78	40.29
23	10.20	11.69	13.09	14.85	32.01	35.17	38.08	41.64
24	10.86	12.40	13.85	15.66	33.20	36.42	39.36	42.98
25	11.52	13.12	14.61	16.47	34.38	37.65	40.65	44.31
26	12.20	13.84	15.38	17.29	35.56	38.89	41.92	45.64
27	12.88	14.57	16.15	18.11	36.74	40.11	43.19	46.96
28	13.56	15.31	16.93	18.94	37.92	41.34	44.46	48.28
29	14.26	16.05	17.71	19.77	39.09	42.56	45.72	49.59
30	14.95	16.79	18.49	20.60	40.26	43.77	46.98	50.89

TABLE B.3 Critical Values for the Wilcoxon Signed Ranks Test Statistics, T

n	$\alpha_{\text{two-tailed}} \leq 0.10$; $\alpha_{\text{one-tailed}} \leq 0.05$	$\alpha_{\text{two-tailed}} \leq 0.05$; $\alpha_{\text{one-tailed}} \leq 0.025$	$\alpha_{\text{two-tailed}} \leq 0.02$; $\alpha_{\text{one-tailed}} \leq 0.01$	$\alpha_{\text{two-tailed}} \leq 0.01$; $\alpha_{\text{one-tailed}} \leq 0.005$
5	0			
6	2	0		
7	3	2	0	
8	5	3	1	0
9	8	5	3	1
10	10	8	5	3
11	13	10	7	5
12	17	13	9	7
13	21	17	12	9
14	25	21	15	12
15	30	25	19	15
16	35	29	23	19
17	41	34	27	23
18	47	40	32	27
19	53	46	37	32
20	60	52	43	37
21	67	58	49	42
22	75	65	55	48

(*Continued*)

TABLE B.3 (*Continued*)

n	$\alpha_{\text{two-tailed}} \leq 0.10$; $\alpha_{\text{one-tailed}} \leq 0.05$	$\alpha_{\text{two-tailed}} \leq 0.05$; $\alpha_{\text{one-tailed}} \leq 0.025$	$\alpha_{\text{two-tailed}} \leq 0.02$; $\alpha_{\text{one-tailed}} \leq 0.01$	$\alpha_{\text{two-tailed}} \leq 0.01$; $\alpha_{\text{one-tailed}} \leq 0.005$
23	83	73	62	54
24	91	81	69	61
25	100	89	76	68
26	110	98	84	75
27	119	107	92	83
28	130	116	101	91
29	140	126	110	100
30	151	137	120	109

Adapted from McCornack, R. L. (1965). Extended tables of the Wilcoxon matched pair signed rank statistic. *Journal of the American Statistical Association, 60*, 864–871. Reprinted with permission from *The Journal of the American Statistical Association*. Copyright 1965 by the American Statistical Association. All rights reserved.

TABLE B.4 Critical Values for the Mann–Whitney *U*-Test Statistic

α	m	1	2	3	4	5	6	7	8	9	10	11	12	13	14	15	16	17	18	19	20
0.10	1																				
	2																				
	3	0	1																		
	4	0	1	3																	
	5	1	2	4	5																
	6	1	3	5	7	9															
	7	1	4	6	8	11	13														
	8	2	5	7	10	13	16	19													
	9	0	2	5	9	12	15	18	22	25											
	10	0	3	6	10	13	17	21	24	28	32										
	11	0	3	7	11	15	19	23	27	31	36	40									
	12	0	4	8	12	17	21	26	30	35	39	44	49								
	13	0	4	9	13	18	23	28	33	38	43	48	53	58							
	14	0	5	10	15	20	25	31	36	41	47	52	58	63	69						
	15	0	5	10	16	22	27	33	39	45	51	57	63	68	74	80					
	16	0	5	11	17	23	29	36	42	48	54	61	67	74	80	86	93				
	17	0	6	12	18	25	31	38	45	52	58	65	72	79	85	92	99	106			
	18	0	6	13	20	27	34	41	48	55	62	69	77	84	91	98	106	113	120		
	19	1	7	14	21	28	36	43	51	58	66	73	81	89	97	104	112	120	128	135	
	20	1	7	15	22	30	38	46	54	62	70	78	86	94	102	110	119	127	135	143	151
0.05	1																				
	2																				
	3			0																	
	4			0	1																
	5		0	1	2	4															
	6		0	2	3	5	7														

APPENDIX B: TABLES OF CRITICAL VALUES 229

TABLE B.4 (*Continued*)

α	m	\multicolumn{20}{c}{n}
		1	2	3	4	5	6	7	8	9	10	11	12	13	14	15	16	17	18	19	20
	7		0	2	4	6	8	11													
	8		1	3	5	8	10	13	15												
	9		1	4	6	9	12	15	18	21											
	10		1	4	7	11	14	17	20	24	27										
	11		1	5	8	12	16	19	23	27	31	34									
	12		2	5	9	13	17	21	26	30	34	38	42								
	13		2	6	10	15	19	24	28	33	37	42	47	51							
	14		3	7	11	16	21	26	31	36	41	46	51	56	61						
	15		3	7	12	18	23	28	33	39	44	50	55	61	66	72					
	16		3	8	14	19	25	30	36	42	48	54	60	65	71	77	83				
	17		3	9	15	20	26	33	39	45	51	57	64	70	77	83	89	96			
	18		4	9	16	22	28	35	41	48	55	61	68	75	82	88	95	102	109		
	19	0	4	10	17	23	30	37	44	51	58	65	72	80	87	94	101	109	116	123	
	20	0	4	11	18	25	32	39	47	54	62	69	77	84	92	100	107	115	123	130	138
0.025	1																				
	2																				
	3																				
	4				0																
	5			0	1	2															
	6			1	2	3	5														
	7			1	3	5	6	8													
	8		0	2	4	6	8	10	13												
	9		0	2	4	7	10	12	15	17											
	10		0	3	5	8	11	14	17	20	23										
	11		0	3	6	9	13	16	19	23	26	30									
	12		1	4	7	11	14	18	22	26	29	33	37								
	13		1	4	8	12	16	20	24	28	33	37	41	45							
	14		1	5	9	13	17	22	26	31	36	40	45	50	55						
	15		1	5	10	14	19	24	29	34	39	44	49	54	59	64					
	16		1	6	11	15	21	26	31	37	42	47	53	59	64	70	75				
	17		2	6	11	17	22	28	34	39	45	51	57	63	69	75	81	87			
	18		2	7	12	18	24	30	36	42	48	55	61	67	74	80	86	93	99		
	19		2	7	13	19	25	32	38	45	52	58	65	72	78	85	92	99	106	113	
	20		2	8	14	20	27	34	41	48	55	62	69	76	83	90	98	105	112	119	127
0.01	1																				
	2																				
	3																				
	4																				
	5				0	1															
	6				1	2	3														
	7			0	1	3	4	6													
	8			0	2	4	6	7	9												
	9			1	3	5	7	9	11	14											
	10			1	3	6	8	11	13	16	19										

(*Continued*)

TABLE B.4 (*Continued*)

α	m	1	2	3	4	5	6	7	8	9	10	11	12	13	14	15	16	17	18	19	20
	11		1	4	7	9	12	15	18	22	25										
	12		2	5	8	11	14	17	21	24	28	31									
	13	0	2	5	9	12	16	20	23	27	31	35	39								
	14	0	2	6	10	13	17	22	26	30	34	38	43	47							
	15	0	3	7	11	15	19	24	28	33	37	42	47	51	56						
	16	0	3	7	12	16	21	26	31	36	41	46	51	56	61	66					
	17	0	4	8	13	18	23	28	33	38	44	49	55	60	66	71	77				
	18	0	4	9	14	19	24	30	36	41	47	53	59	65	70	76	82	88			
	19	1	4	9	15	20	26	32	38	44	50	56	63	69	75	82	88	94	101		
	20	1	5	10	16	22	28	34	40	47	53	60	67	73	80	87	93	100	107	114	

Adapted from Milton, R. C. (1964). An extended table of critical values for the Mann–Whitney (Wilcoxon) two-sample statistic. *Journal of the American Statistical Association, 59*, 925–934. Reprinted with permission from *The Journal of the American Statistical Association*. Copyright 1964 by the American Statistical Association. All rights reserved.

TABLE B.5 Critical Values for the Friedman Test Statistic, F_r

k	N	$\alpha \leq 0.10$	$\alpha \leq 0.05$	$\alpha \leq 0.025$	$\alpha \leq 0.01$
3	3	6.000	6.000		
	4	6.000	6.500	8.000	8.000
	5	5.200	6.400	7.600	8.400
	6	5.333	7.000	8.333	9.000
	7	5.429	7.143	7.714	8.857
	8	5.250	6.250	7.750	9.000
	9	5.556	6.222	8.000	8.667
	10	5.000	6.200	7.800	9.600
	11	4.909	6.545	7.818	9.455
	12	5.167	6.500	8.000	9.500
	13	4.769	6.000	7.538	9.385
	14	5.143	6.143	7.429	9.000
	15	4.933	6.400	7.600	8.933
4	2	6.000	6.000		
	3	6.600	7.400	8.200	9.000
	4	6.300	7.800	8.400	9.600
	5	6.360	7.800	8.760	9.960
	6	6.400	7.600	8.800	10.200
	7	6.429	7.800	9.000	10.371
	8	6.300	7.650	9.000	10.500
	9	6.467	7.800	9.133	10.867
	10	6.360	7.800	9.120	10.800
	11	6.382	7.909	9.327	11.073

APPENDIX B: TABLES OF CRITICAL VALUES

TABLE B.5 (*Continued*)

k	N	$\alpha \leq 0.10$	$\alpha \leq 0.05$	$\alpha \leq 0.025$	$\alpha \leq 0.01$
	12	6.400	7.900	9.200	11.100
	13	6.415	7.985	7.369	11.123
	14	6.343	7.886	9.343	11.143
	15	6.440	8.040	9.400	11.240
5	2	7.200	7.600	8.000	8.000
	3	7.467	8.533	9.600	10.133
	4	7.600	8.800	9.800	11.200
	5	7.680	8.960	10.240	11.680
	6	7.733	9.067	10.400	11.867
	7	7.771	9.143	10.514	12.114
	8	7.800	9.300	10.600	12.300
	9	7.733	9.244	10.667	12.444
	10	7.760	9.280	10.720	12.480
6	2	8.286	9.143	9.429	9.714
	3	8.714	9.857	10.810	11.762
	4	9.000	10.286	11.429	12.714
	5	9.000	10.486	11.743	13.229
	6	9.048	10.571	12.000	13.619
	7	9.122	10.674	12.061	13.857
	8	9.143	10.714	12.214	14.000
	9	9.127	10.778	12.302	14.143
	10	9.143	10.800	12.343	14.229

Adapted from Martin, L., Leblanc, R., & Toan N. K. (1993). Tables for the Friedman rank test. *The Canadian Journal of Statistics/La Revue Canadienne de Statistique, 21*(1), 39–43. Reprinted with permission from *The Canadian Journal of Statistics*. Copyright 1993 by the Statistical Society of Canada. All rights reserved.

TABLE B.6 The Critical Values for the Kruskal–Wallis *H*-Test Statistic

			$k=3$		
n_1	n_2	n_3	$\alpha \leq 0.10$	$\alpha \leq 0.05$	$\alpha \leq 0.01$
2	2	2	4.571429	–	–
3	1	1	–	–	–
3	2	1	4.285714	–	–
3	2	2	4.464286	4.714286	–
3	3	1	4.571429	5.142857	–
3	3	2	4.555556	5.361111	–
3	3	3	4.622222	5.600000	6.488889
4	2	1	4.500000	–	–
4	2	2	4.458333	5.333333	–

(*Continued*)

TABLE B.6 (*Continued*)

			$k=3$		
n_1	n_2	n_3	$\alpha \leq 0.10$	$\alpha \leq 0.05$	$\alpha \leq 0.01$
4	3	1	4.055556	5.208333	–
4	3	2	4.511111	5.444444	6.444444
4	3	3	4.700000	5.790909	6.745455
4	4	1	4.166667	4.966667	6.666667
4	4	2	4.554545	5.454545	7.036364
4	4	3	4.545455	5.598485	7.143939
4	4	4	4.653846	5.692308	7.653846
5	2	1	4.200000	5.000000	–
5	2	2	4.373333	5.160000	6.533333
5	3	1	4.017778	4.871111	–
5	3	2	4.650909	5.250909	6.821818
5	3	3	4.533333	5.648485	7.078788
5	4	1	3.987273	4.985455	6.954545
5	4	2	4.540909	5.272727	7.204545
5	4	3	4.548718	5.656410	7.444872
5	4	4	4.668132	5.657143	7.760440
5	5	1	4.109091	5.127273	7.309091
5	5	2	4.623077	5.338462	7.338462
5	5	3	4.545055	5.626374	7.578022
5	5	4	4.522857	5.665714	7.791429
5	5	5	4.560000	5.780000	8.000000
6	2	1	4.200000	4.822222	–
6	2	2	4.436364	5.345455	6.654545
6	3	1	3.909091	4.854545	6.581818
6	3	2	4.681818	5.348485	6.969697
6	3	3	4.538462	5.615385	7.192308
6	4	1	4.037879	4.946970	7.083333
6	4	2	4.493590	5.262821	7.339744
6	4	3	4.604396	5.604396	7.467033
6	4	4	4.523810	5.666667	7.795238
6	5	1	4.128205	4.989744	7.182051
6	5	2	4.595604	5.318681	7.375824
6	5	3	4.535238	5.601905	7.590476
6	5	4	4.522500	5.660833	7.935833
6	5	5	4.547059	5.698529	8.027941
6	6	1	4.000000	4.857143	7.065934
6	6	2	4.438095	5.409524	7.466667
6	6	3	4.558333	5.625000	7.725000
6	6	4	4.547794	5.724265	8.000000
6	6	5	4.542484	5.764706	8.118954
6	6	6	4.538012	5.719298	8.222222
7	1	1	4.266667	–	–
7	2	1	4.200000	4.706494	–
7	2	2	4.525974	5.142857	7.000000
7	3	1	4.173160	4.952381	6.649351

APPENDIX B: TABLES OF CRITICAL VALUES

TABLE B.6 (*Continued*)

			$k = 3$		
n_1	n_2	n_3	$\alpha \leq 0.10$	$\alpha \leq 0.05$	$\alpha \leq 0.01$
7	3	2	4.582418	5.357143	6.838828
7	3	3	4.602826	5.620094	7.227630
7	4	1	4.120879	4.986264	6.986264
7	4	2	4.549451	5.375981	7.304553
7	4	3	4.527211	5.623129	7.498639
7	4	4	4.562500	5.650000	7.814286
7	5	1	4.035165	5.063736	7.060597
7	5	2	4.484898	5.392653	7.449796
7	5	3	4.535238	5.588571	7.697143
7	5	4	4.541597	5.732773	7.931092
7	5	5	4.540056	5.707563	8.100840
7	6	1	4.032653	5.066667	7.254422
7	6	2	4.500000	5.357143	7.490476
7	6	3	4.550420	5.672269	7.756303
7	6	4	4.561625	5.705882	8.016340
7	6	5	4.559733	5.769925	8.156725
7	6	6	4.530075	5.730075	8.257143
7	7	1	3.985714	4.985714	7.157143
7	7	2	4.490546	5.398109	7.490546
7	7	3	4.590103	5.676937	7.809524
7	7	4	4.558897	5.765664	8.141604
7	7	5	4.545564	5.745564	8.244812
7	7	6	4.568027	5.792517	8.341497
7	7	7	4.593692	5.818182	8.378479
8	1	1	4.418182	–	–
8	2	1	4.011364	4.909091	–
8	2	2	4.586538	5.355769	6.663462
8	3	1	4.009615	4.881410	6.804487
8	3	2	4.450549	5.315934	6.986264
8	3	3	4.504762	5.616667	7.254762
8	4	1	4.038462	5.043956	6.972527
8	4	2	4.500000	5.392857	7.350000
8	4	3	4.529167	5.622917	7.585417
8	4	4	4.560662	5.779412	7.852941
8	5	1	3.967143	4.868571	7.110000
8	5	2	4.466250	5.415000	7.440000
8	5	3	4.514338	5.614338	7.705515
8	5	4	4.549020	5.717647	7.992157
8	5	5	4.555263	5.769298	8.115789
8	6	1	4.014583	5.014583	7.256250
8	6	2	4.441176	5.404412	7.522059
8	6	3	4.573529	5.678105	7.795752
8	6	4	4.562865	5.742690	8.045322
8	6	5	4.550263	5.750263	8.210263

(*Continued*)

TABLE B.6 (*Continued*)

			$k=3$		
n_1	n_2	n_3	$\alpha \leq 0.10$	$\alpha \leq 0.05$	$\alpha \leq 0.01$
8	6	6	4.598810	5.770238	8.294048
8	7	1	4.045431	5.041229	7.307773
8	7	2	4.450980	5.403361	7.571429
8	7	3	4.555556	5.698413	7.827068
8	7	4	4.548496	5.759211	8.118045
8	7	5	4.550612	5.777449	8.241939
8	7	6	4.552876	5.781231	8.332715
8	7	7	4.573687	5.795031	8.356296
8	8	1	4.044118	5.039216	7.313725
8	8	2	4.508772	5.407895	7.653509
8	8	3	4.555263	5.734211	7.889474
8	8	4	4.578571	5.742857	8.167857
8	8	5	4.572727	5.761039	8.297403
8	8	6	4.572134	5.778656	8.366601
8	8	7	4.570652	5.791149	8.418866
8	8	8	4.595000	5.805000	8.465000
9	1	1	4.545455	–	–
9	2	1	3.905983	4.841880	6.346154
9	2	2	4.483516	5.260073	6.897436
9	3	1	4.073260	4.952381	6.886447
9	3	2	4.492063	5.339683	6.990476
9	3	3	4.633333	5.588889	7.355556
9	4	1	3.971429	5.071429	7.171429
9	4	2	4.488889	5.400000	7.363889
9	4	3	4.514706	5.651961	7.613971
9	4	4	4.576253	5.703704	7.909586
9	5	1	4.055556	5.040000	7.148889
9	5	2	4.464706	5.395588	7.447059
9	5	3	4.587364	5.669717	7.733333
9	5	4	4.531384	5.712671	8.024561
9	5	5	4.557193	5.769825	8.169825
9	6	1	3.933824	5.049020	7.247549
9	6	2	4.481481	5.392157	7.494553
9	6	3	4.541910	5.664717	7.822612
9	6	4	4.545614	5.744737	8.108772
9	6	5	4.573651	5.761905	8.230794
9	6	6	4.554113	5.808081	8.307359
9	7	1	4.011204	5.042017	7.270464
9	7	2	4.480089	5.429128	7.636591
9	7	3	4.535338	5.656140	7.860652
9	7	4	4.547732	5.731406	8.131406
9	7	5	4.565492	5.757988	8.287941
9	7	6	4.570864	5.782985	8.353284
9	7	7	4.583851	5.802622	8.403037
9	8	1	3.986355	4.984893	7.394250

TABLE B.6 (*Continued*)

			$k = 3$		
n_1	n_2	n_3	$\alpha \leq 0.10$	$\alpha \leq 0.05$	$\alpha \leq 0.01$
9	8	2	4.491667	5.419737	7.642105
9	8	3	4.568651	5.717460	7.927381
9	8	4	4.559163	5.744228	8.203102
9	8	5	4.551252	5.783465	8.318050
9	8	6	4.560688	5.775362	8.408514
9	8	7	4.563770	5.807579	8.450000
9	8	8	4.582821	5.809744	8.494359
9	9	1	4.007018	4.961404	7.333333
9	9	2	4.460317	5.411111	7.692063
9	9	3	4.565657	5.708514	7.959596
9	9	4	4.550066	5.751647	8.202240
9	9	5	4.587440	5.770048	8.370048
9	9	6	4.555556	5.814444	8.427778
9	9	7	4.567326	5.802198	8.468864
9	9	8	4.570750	5.815052	8.514720
9	9	9	4.582011	5.844797	8.564374
10	1	1	4.653846	4.653846	–
10	2	1	4.114286	4.839560	6.428571
10	2	2	4.434286	5.120000	6.537143
10	3	1	3.996190	5.076190	6.851429
10	3	2	4.470000	5.361667	7.041667
10	3	3	4.529412	5.588235	7.360294
10	4	1	4.042500	5.017500	7.105000
10	4	2	4.462500	5.344853	7.356618
10	4	3	4.587582	5.654248	7.616993
10	4	4	4.564912	5.715789	7.907018
10	5	1	3.988235	4.905882	7.107353
10	5	2	4.454902	5.388235	7.513725
10	5	3	4.552047	5.618713	7.752047
10	5	4	4.556842	5.744211	8.047895
10	5	5	4.574286	5.777143	8.162857
10	6	1	3.967320	5.041830	7.316340
10	6	2	4.479532	5.405848	7.588304
10	6	3	4.542105	5.655789	7.882105
10	6	4	4.550476	5.726190	8.142857
10	6	5	4.554978	5.754978	8.267532
10	6	6	4.575494	5.780237	8.338340
10	7	1	3.981454	4.985965	7.252130
10	7	2	4.491880	5.377444	7.641203
10	7	3	4.545034	5.698095	7.901224
10	7	4	4.550278	5.751206	8.172356
10	7	5	4.567250	5.763862	8.295652
10	7	6	4.563043	5.798758	8.376915
10	7	7	4.562286	5.796571	8.419429

(*Continued*)

TABLE B.6 (*Continued*)

			$k=3$		
n_1	n_2	n_3	$\alpha \leq 0.10$	$\alpha \leq 0.05$	$\alpha \leq 0.01$
10	8	1	3.963947	5.037632	7.358684
10	8	2	4.482857	5.429286	7.720714
10	8	3	4.533983	5.711688	7.977273
10	8	4	4.550988	5.744466	8.206126
10	9	5	4.556522	5.789130	8.344022
10	9	6	4.573333	5.793833	8.397833
10	9	7	4.564484	5.810637	8.480967
10	9	8	4.561538	5.829060	8.494017
10	9	1	4.025714	4.988571	7.436508
10	9	2	4.476479	5.446176	7.693795
10	9	3	4.570751	5.700659	7.997628
10	9	4	4.556401	5.757609	8.223430
10	9	5	4.547556	5.792000	8.380222
10	9	6	4.561231	5.813128	8.449436
10	9	7	4.559707	5.817610	8.507475
10	9	8	4.567063	5.833730	8.544489
10	9	9	4.578982	5.830706	8.575698
10	10	1	3.987013	5.054545	7.501299
10	10	2	4.477470	5.449802	7.726482
10	10	3	4.559420	5.687681	8.026087
10	10	4	4.567000	5.776000	8.263000
10	10	5	4.554462	5.793231	8.403692
10	10	6	4.561823	5.796011	8.472934
10	10	7	4.558277	5.820408	8.536508
10	10	8	4.565025	5.837438	8.565887
10	10	9	4.567050	5.837241	8.606130
10	10	10	4.583226	5.855484	8.640000

				$k=4$		
n_1	n_2	n_3	n_4	$\alpha \leq 0.10$	$\alpha \leq 0.05$	$\alpha \leq 0.01$
2	2	2	1	5.357143	5.678571	–
2	2	2	2	5.666667	6.166667	6.666667
3	2	1	1	4.892857	–	–
3	2	2	1	5.555556	5.833333	–
3	2	2	2	5.644444	6.333333	7.133333
3	3	1	1	5.333333	6.333333	–
3	3	2	1	5.622222	6.244444	7.044444
3	3	2	2	5.745455	6.527273	7.636364
3	3	3	1	5.654545	6.600000	7.400000
3	3	3	2	5.878788	6.727273	8.015152
3	3	3	3	5.974359	6.897436	8.435897
4	2	1	1	5.250000	5.833333	–
4	2	2	1	5.533333	6.133333	7.000000
4	2	2	2	5.754545	6.545455	7.390909

APPENDIX B: TABLES OF CRITICAL VALUES

TABLE B.6 (*Continued*)

				$k=4$		
n_1	n_2	n_3	n_4	$\alpha \leq 0.10$	$\alpha \leq 0.05$	$\alpha \leq 0.01$
4	3	1	1	5.066667	6.177778	7.066667
4	3	2	1	5.572727	6.309091	7.454545
4	3	2	2	5.750000	6.621212	7.871212
4	3	3	1	5.666667	6.545455	7.757576
4	3	3	2	5.858974	6.782051	8.320513
4	3	3	3	6.000000	6.967033	8.653846
4	4	1	1	5.181818	5.945455	7.909091
4	4	2	1	5.568182	6.386364	7.909091
4	4	2	2	5.807692	6.730769	8.346154
4	4	3	1	5.660256	6.634615	8.217949
4	4	3	2	5.901099	6.873626	8.620879
4	4	3	3	6.004762	7.038095	8.866667
4	4	4	1	5.653846	6.725275	8.587912
4	4	4	2	5.914286	6.957143	8.871429
4	4	4	3	6.029167	7.129167	9.075000
4	4	4	4	6.088235	7.235294	9.286765
5	1	1	1	5.333333	–	–
5	2	1	1	5.266667	5.960000	–
5	2	2	1	5.541818	6.109091	7.276364
5	2	2	2	5.636364	6.563636	7.772727
5	3	1	1	5.130909	6.003636	7.400000
5	3	2	1	5.518182	6.363636	7.757576
5	3	2	2	5.771795	6.664103	8.202564
5	3	3	1	5.656410	6.641026	8.117949
5	3	3	2	5.865934	6.821978	8.606593
5	3	3	3	6.020952	7.011429	8.840000
5	4	1	1	5.254545	6.040909	7.909091
5	4	2	1	5.580769	6.419231	8.173077
5	4	2	2	5.782418	6.725275	8.472527
5	4	3	1	5.639560	6.681319	8.408791
5	4	3	2	5.901905	6.925714	8.801905
5	4	3	3	6.029167	7.093333	9.029167
5	4	4	1	5.674286	6.760000	8.725714
5	4	4	2	5.947500	6.990000	9.002500
5	4	4	3	6.035294	7.172794	9.220588
5	4	4	4	6.066667	7.262745	9.392157
5	5	1	1	5.153846	6.076923	8.107692
5	5	2	1	5.564835	6.540659	8.327473
5	5	2	2	5.794286	6.777143	8.634286
5	5	3	1	5.662857	6.737143	8.611429
5	5	3	2	5.921667	6.946667	8.946667
5	5	3	3	6.023529	7.117647	9.188235
5	5	4	1	5.670000	6.782500	8.870000
5	5	4	2	5.944853	7.032353	9.156618

(*Continued*)

TABLE B.6 (*Continued*)

				$k = 4$		
n_1	n_2	n_3	n_4	$\alpha \leq 0.10$	$\alpha \leq 0.05$	$\alpha \leq 0.01$
5	5	4	3	6.052288	7.217647	9.356863
5	5	4	4	6.070175	7.291228	9.536842
5	5	5	1	5.682353	6.829412	9.052941
5	5	5	2	5.945098	7.074510	9.286275
5	5	5	3	6.043275	7.250292	9.495906
5	5	5	4	6.082105	7.327895	9.669474
5	5	5	5	6.097143	7.377143	9.800000

Adapted from Meyer, J. P., & Seaman, M. A. (2008, March). A comparison of the exact Kruskal–Wallis distribution to asymptotic approximations for $N \leq 105$. Paper presented at the Annual Meeting of the American Educational Research Association, New York. Reprinted with permission of the authors.

TABLE B.7 Critical Values for the Spearman Rank-Order Correlation Coefficient, r_s

n	$\alpha_{\text{two-tailed}} \leq 0.10$; $\alpha_{\text{one-tailed}} \leq 0.05$	$\alpha_{\text{two-tailed}} \leq 0.05$; $\alpha_{\text{one-tailed}} \leq 0.025$	$\alpha_{\text{two-tailed}} \leq 0.02$; $\alpha_{\text{one-tailed}} \leq 0.01$	$\alpha_{\text{two-tailed}} \leq 0.01$; $\alpha_{\text{one-tailed}} \leq 0.005$
4	1.000			
5	0.900	1.000	1.000	
6	0.829	0.886	0.943	1.000
7	0.714	0.786	0.893	0.929
8	0.643	0.738	0.833	0.881
9	0.600	0.700	0.783	0.833
10	0.564	0.648	0.745	0.794
11	0.536	0.618	0.709	0.755
12	0.503	0.587	0.671	0.727
13	0.484	0.560	0.648	0.703
14	0.464	0.538	0.622	0.675
15	0.443	0.521	0.604	0.654
16	0.429	0.503	0.582	0.635
17	0.414	0.485	0.566	0.615
18	0.401	0.472	0.550	0.600
19	0.391	0.460	0.535	0.584
20	0.380	0.447	0.520	0.570
21	0.370	0.435	0.508	0.556
22	0.361	0.425	0.496	0.544
23	0.353	0.415	0.486	0.532
24	0.344	0.406	0.476	0.321
25	0.337	0.398	0.466	0.511
26	0.331	0.390	0.457	0.501
27	0.324	0.382	0.448	0.491
28	0.317	0.375	0.440	0.483
29	0.312	0.368	0.433	0.475

TABLE B.7 (*Continued*)

n	$\alpha_{\text{two-tailed}} \leq 0.10$; $\alpha_{\text{one-tailed}} \leq 0.05$	$\alpha_{\text{two-tailed}} \leq 0.05$; $\alpha_{\text{one-tailed}} \leq 0.025$	$\alpha_{\text{two-tailed}} \leq 0.02$; $\alpha_{\text{one-tailed}} \leq 0.01$	$\alpha_{\text{two-tailed}} \leq 0.01$; $\alpha_{\text{one-tailed}} \leq 0.005$
30	0.306	0.362	0.425	0.467
31	0.301	0.356	0.418	0.459
32	0.296	0.350	0.412	0.452
33	0.291	0.345	0.405	0.446
34	0.287	0.340	0.399	0.439
35	0.283	0.335	0.394	0.433
36	0.279	0.330	0.388	0.427
37	0.275	0.325	0.383	0.421
38	0.271	0.321	0.378	0.415
39	0.267	0.317	0.373	0.410
40	0.264	0.313	0.368	0.405
41	0.261	0.309	0.364	0.400
42	0.257	0.305	0.359	0.395
43	0.254	0.301	0.355	0.391
44	0.251	0.298	0.351	0.386
45	0.248	0.294	0.347	0.382
46	0.246	0.291	0.343	0.378
47	0.243	0.288	0.340	0.374
48	0.240	0.285	0.336	0.370
49	0.238	0.282	0.333	0.366
50	0.235	0.279	0.329	0.363

Adapted from Zar, J. H. (1972). Significance testing of the Spearman rank correlation coefficient. *Journal of the American Statistical Association, 67*, 578–580. Reprinted with permission from *The Journal of the American Statistical Association*. Copyright 1972 by the American Statistical Association. All rights reserved.

TABLE B.8 Critical Values for the Pearson Product-Moment Correlation Coefficient, *r*

df	$\alpha_{\text{two-tailed}} \leq 0.10$; $\alpha_{\text{one-tailed}} \leq 0.05$	$\alpha_{\text{two-tailed}} \leq 0.05$; $\alpha_{\text{one-tailed}} \leq 0.025$	$\alpha_{\text{two-tailed}} \leq 0.025$; $\alpha_{\text{one-tailed}} \leq 0.0125$	$\alpha_{\text{two-tailed}} \leq 0.01$; $\alpha_{\text{one-tailed}} \leq 0.005$
1	0.988	0.997	0.999	0.999
2	0.900	0.950	0.975	0.990
3	0.805	0.878	0.924	0.959
4	0.729	0.811	0.868	0.917
5	0.669	0.754	0.817	0.875
6	0.621	0.707	0.771	0.834
7	0.582	0.666	0.732	0.798
8	0.549	0.632	0.697	0.765
9	0.521	0.602	0.667	0.735
10	0.497	0.576	0.640	0.708
11	0.476	0.553	0.616	0.684

(*Continued*)

TABLE B.8 (*Continued*)

df	$\alpha_{\text{two-tailed}} \leq 0.10$; $\alpha_{\text{one-tailed}} \leq 0.05$	$\alpha_{\text{two-tailed}} \leq 0.05$; $\alpha_{\text{one-tailed}} \leq 0.025$	$\alpha_{\text{two-tailed}} \leq 0.025$; $\alpha_{\text{one-tailed}} \leq 0.0125$	$\alpha_{\text{two-tailed}} \leq 0.01$; $\alpha_{\text{one-tailed}} \leq 0.005$
12	0.458	0.532	0.594	0.661
13	0.441	0.514	0.575	0.641
14	0.426	0.497	0.557	0.623
15	0.412	0.482	0.541	0.606
16	0.400	0.468	0.526	0.590
17	0.389	0.456	0.512	0.575
18	0.378	0.444	0.499	0.561
19	0.369	0.433	0.487	0.549
20	0.360	0.423	0.476	0.537
21	0.352	0.413	0.466	0.526
22	0.344	0.404	0.456	0.515
23	0.337	0.396	0.447	0.505
24	0.330	0.388	0.439	0.496
25	0.323	0.381	0.430	0.487
26	0.317	0.374	0.423	0.479
27	0.311	0.367	0.415	0.471
28	0.306	0.361	0.409	0.463
29	0.301	0.355	0.402	0.456
30	0.296	0.349	0.396	0.449
31	0.291	0.344	0.390	0.442
32	0.287	0.339	0.384	0.436
33	0.283	0.334	0.378	0.430
34	0.279	0.329	0.373	0.424
35	0.275	0.325	0.368	0.418
36	0.271	0.320	0.363	0.413
37	0.267	0.316	0.359	0.408
38	0.264	0.312	0.354	0.403
39	0.260	0.308	0.350	0.398
40	0.257	0.304	0.346	0.393
41	0.254	0.301	0.342	0.389
42	0.251	0.297	0.338	0.384
43	0.248	0.294	0.334	0.380
44	0.246	0.291	0.330	0.376
45	0.243	0.288	0.327	0.372
46	0.240	0.285	0.323	0.368
47	0.238	0.282	0.320	0.365
48	0.235	0.279	0.317	0.361
49	0.233	0.276	0.314	0.358
50	0.231	0.273	0.311	0.354

APPENDIX B: TABLES OF CRITICAL VALUES

TABLE B.9 Factorials

n	$n!$
1	1
2	2
3	6
4	24
5	120
6	720
7	5040
8	40320
9	362880
10	3628800
11	39916800
12	479001600
13	6227020800
14	87178291200
15	1307674368000
16	20922789888000
17	355687428096000
18	6402373705728000
19	121645100408832000
20	2432902008176640000
21	51090942171709440000
22	1124000727777607680000
23	25852016738884976640000
24	620448401733239439360000
25	15511210043330985984000000

TABLE B.10 Critical Values for the Runs Test for Randomness

	One-tailed Alternative; $\alpha = 0.05$											
						n_2						
n_1	2	3	4	5	6	7	8	9	10	11	12
2	–	–	–	–	–	–	2	2	2	2	2
	–	–	–	–	–	–	–	–	–	–	–
3	–	–	–	2	2	2	2	2	3	3	3
	–	–	7	–	–	–	–	–	–	–	–
4	–	–	2	2	3	3	3	3	3	3	4
	–	7	8	9	9	9	–	–	–	–	–
5	–	2	2	3	3	3	3	4	4	4	4
	–	–	9	9	10	10	11	11	11	–	–

(*Continued*)

TABLE B.10 (*Continued*)

	One-tailed Alternative; $\alpha = 0.05$										
					n_2						
n_1	2	3	4	5	6	7	8	9	10	11	12
6	–	2	3	3	3	4	4	4	5	5	5
	–	–	9	10	11	11	12	12	12	13	13
7	–	2	3	3	4	4	4	5	5	5	6
	–	–	9	10	11	12	13	13	13	14	14
8	2	2	3	3	4	4	5	5	6	6	6
	–	–	–	11	12	13	13	14	14	15	15
9	2	2	3	4	4	5	5	6	6	6	7
	–	–	–	11	12	13	14	14	15	15	16
10	2	3	3	4	5	5	6	6	6	7	7
	–	–	–	11	12	13	14	15	16	16	17
11	2	3	3	4	5	5	6	6	7	7	8
	–	–	–	–	13	14	15	15	16	17	17
12	2	3	4	4	5	6	6	7	7	8	8
	–	–	–	–	13	14	15	16	17	17	18

	One-tailed Alternative; $\alpha = 0.025$										
2	–	–	–	–	–	–	–	–	–	–	2
	–	–	–	–	–	–	–	–	–	–	–
3	–	–	–	–	2	2	2	2	2	2	2
	–	–	–	–	–	–	–	–	–	–	–
4	–	–	–	2	2	2	3	3	3	3	3
	–	–	–	9	9	–	–	–	–	–	–
5	–	–	2	2	3	3	3	3	3	4	4
	–	–	9	10	10	11	11	–	–	–	–
6	–	2	2	3	3	3	3	4	4	4	4
	–	–	9	10	11	12	12	13	13	13	13
7	–	2	2	3	3	3	4	4	5	5	5
	–	–	–	11	12	13	13	14	14	14	14
8	–	2	3	3	3	4	4	5	5	5	6
	–	–	–	11	12	13	14	14	15	15	16
9	–	2	3	3	4	4	5	5	5	6	6
	–	–	–	–	13	14	14	15	16	16	16
10	–	2	3	3	4	5	5	5	6	6	7
	–	–	–	–	13	14	15	16	16	17	17
11	–	2	3	4	4	5	5	6	6	7	7
	–	–	–	–	13	14	15	16	17	17	18
12	2	2	3	4	4	5	6	6	7	7	7
	–	–	–	–	13	14	16	16	17	18	19

Adapted from Table D.5 and D.6 of Janke, S. J., & Tinsley, F. C. (2005). *Introduction to linear models and statistical inference*. Hoboken, NJ: John Wiley & Sons, Inc. Reprinted with permission of John Wiley & Sons, Inc. Copyright 2005 by John Wiley & Sons, Inc. All rights reserved.

BIBLIOGRAPHY

American Psychological Association. (2001). *Publication manual of the American Psychological Association*, (5th ed.). Washington, DC: Author.

Anderson, N. H. (1961). Scales and statistics: Parametric and nonparametric. *Psychological Bulletin*, 58(4), 305–316.

Cohen, J. (1988). *Statistical power analysis for the behavioral sciences*. 2nd ed. New York: Academic Press.

Cohen, J. (1992). A power primer. *Psychological Bulletin*, 112(1), 155–159.

Daniel, W. W. (1990). *Applied nonparametric statistics*. 2nd ed. Boston: PWS-KENT Publishing Company.

Field, A. (2005). *Discovering statistics using SPSS*. 2nd ed. London: Sage Publications.

Gaito, J. (1980). Measurement scales and statistics: Resurgence of an old misconception. *Psychological Bulletin*, 87(3), 564–567.

Garrett, H. E. (1966). *Statistics in psychology and education*. 6th ed. New York: David McKay Company, Inc.

Guilford, J. P. (1956). *Fundamental statistics in psychology and education*. 4th ed. New York: McGraw-Hill.

Hastings, C. (1955). *Approximations for digital computers*. Princeton, NJ: Princeton University Press.

Malthouse, E. (2001). Methodological and statistical concerns of the experimental behavioral researcher. *Journal of Consumer Psychology*, 10(1/2), 111–112.

Nanna, M. J., and Sawilowsky, S. S. (1998). Analysis of Likert scale data in disability and medical rehabilitation research. *Psychological Methods*, 3(1), 55–67.

Osborne, J. W., and Overbay, A. (2004). The power of outliers (and why researchers should always check for them). *Practical Assessment, Research & Evaluation*, 9(6). Retrieved January 31, 2008 from http://PAREonline.net/getvn.asp?v=9&n=6.

Pett, M. A. (1997). *Nonparametric statistics for health care research: Statistics for small samples and unusual distributions.* Thousand Oaks, CA: Sage Publications.

Salkind, N. J. (2004). *Statistics for people who (think they) hate statistics.* 2nd ed. Thousand Oaks, CA: Sage Publications.

Siegel, S., and Castellan, N. J. (1988). *Nonparametric statistics for the behavioral sciences.* 2nd ed. New York: McGraw-Hill.

Smirnov, N. V. (1948). Table for estimating the goodness of fit of empirical distributions. *Annals of Mathematical Statistics*, 19, 279–281.

Stake, R. E. (1960). Review of elementary statistics by P. G. Hoel. *Educational and Psychological Measurement*, 20, 871–873.

Stevens, S. S. (1946). On the theory of scales of measurement. *Science*, 103, 677–680.

Townsend, J. T., and Ashby, F.G. (1984). Measurement scales and statistics: The misconception misconceived. *Psychological Bulletin*, 96(2), 394–401.

Velleman, P. F., and Wilkinson, L. (1993). Nominal, ordinal, interval, and ratio typologies are misleading. *The American Statistician*, 47(1), 65–72.

Warner, R. M. (2008). *Applied statistics: From bivariate through multivariate techniques.* Thousand Oaks, CA: Sage Publications.

INDEX

Alpha, α, (See Type I error)
Alternate hypothesis, 4

Beta, β, (See Type II error)
Biserial correlation, 4, 134
   small samples, 146-149
   using SPSS, 149
Bonferroni correction procedure, 81, 101

Categorical data, 3
Chi-square distribution table, 226–227
Chi-square goodness-of-fit test, 4, 156
   category frequencies equal, 157–160
   category frequencies not equal, 160–163
   computing, 156
   using SPSS, 163–167
Chi-square test of independence, 4, 167, 169–173
   computing, 168–169
   using SPSS, 174–179
Contingency tables, 2 × 2, 180
Correlation coefficient, 123
   of a dichotomous variable and a rank-order variable, 136
   of a dichotomous variable and an interval scale variable, 134–135
   biserial, 4, 134
   point-biserial, 4, 134
   Spearman rank-order, 4, 122–123
Counts, 8–9
Cramer's V, 169, 173
Critical value, 6
   tables of, 219–242
Cumulative frequency distributions, 26

Dichotomous scale
   continuous, 3
   discrete, 3
Divergence, 27

Effect size, 39–40, 59, 169
   Cramer's V, 169, 173
   Phi coefficient, 169, 184

Factorials, table of, 241
Fisher exact test, 4, 179, 180–184
    computing, 180
    using SPSS, 184–185
Friedman test, 4, 79–80
    computing, 80–81
    critical values table for, 230–231
    large sample approximation, 81
    post hoc test for, 81
    sample contrasts for, 81
    small samples with ties, 80
    small samples without ties, 80
    using SPSS, 88–90

Histogram, 14–15
Homogeneity of variance, 14

Interval scale, 3

Kolmogorov-Smirnov one-sample test, 26
    computing, 26–28
    using SPSS, 32–35
Kruskal-Wallis H test, 4, 99–100
    computing, 100–101
    correction for ties, 100–101
    critical values table for, 231–238
    large sample approximation, 100
    post hoc test for, 101
    sample contrasts for, 101
    small data samples, 100
    using SPSS, 106–110
Kurtosis, 16–19
    computing, 18
    leptokurtosis, 16–17
    platykurtosis, 17
    standard error of, 18
    using SPSS, 23–26

Leptokurtosis, 16–17
Likert scale, 5

Mann-Whitney U test, 4, 57–58
    computing, 58–59
    confidence interval for, 66–67
    critical values table for, 228–230
    large sample approximation, 58–59
    small samples, 58
    using SPSS, 62–66
Mean, 16

Measurement scale, 3
Median, 16
Mode, 16

Nominal data, 3
Nonparametric tests, comparison with
        parametric tests, 4
Normal curve, 15–16
    properties of, 16
Normal distribution, 13–16
Normal distribution table, 219–226
Normality
    assumptions of, 1–2
    measures of, 13
Null hypothesis, 4

Observed frequency distribution, 26–27
Obtained value, 4, 6
Ordinal scale, 3
Ordinate of the normal curve, 16
    computing, 135
    table of, 219–226
Outliers, 2

Parametric tests, 1–2
    comparison with nonparametric
        tests, 4
Pearson product-moment correlation, 134
    critical values table for, 239–240
Phi coefficient, 169, 184
Platykurtosis, 17
Point-biserial correlation, 4, 134
    large samples approximation, 142–146
    small data samples, 136–139
    using SPSS, 139–142
Post hoc comparisons, 81, 101

Randomness, see runs test
Ranking data, 6–7
    with tied values, 7–8
Rank-order scale, 3
Ratio scale, 3
Relative empirical frequency distribution,
        26–27
Relative observed frequency distribution,
        26–27
Runs test, 4, 192
    computing, 193–194
    critical values table for, 241–242
    large sample approximation, 200–202

INDEX

referencing a custom value, 202–204
referencing a custom value using SPSS, 204–208
small data samples, 194–195
using SPSS, 195–200

Sample size, 2
Sampling, 13
Sample contrasts, 81, 101
Scale, see measurement scale
Skewness, 16–19
   computing, 18
   standard error of, 18
   using SPSS, 23–26
Spearman rank-order correlation, 4, 122–123
   computing, 124
   computing Student's t, 125
   computing z, 124–125
   critical values table for, 238–239
   small data sample with ties, 128–131
   small data sample without ties, 125–128
   using SPSS, 131–133
SPSS, 212–218
Standard deviation, 14
Standard error of

kurtosis, 18
skewness, 18
Symmetry, 16

Transformation, 2
Type I error, 5, 81, 101
Type II error, 5

Variance, 13–14
   homogeneity of, 14

Wald-Wolfowitz test, 192
Wilcoxon rank sum test, 57, 66
Wilcoxon signed ranks test, 4, 38–39
   computing, 39–40
   confidence interval for, 45–47
   critical values table for, 227–228
   large sample approximation, 39
   small samples, 39
   using SPSS, 42–45
Wilcoxon W, 66

Yates's continuity correction, 169

z-score, 14–15,
   table of, 219–226